普通高等教育"十三五"规划教材
高等院校计算机系列教材

C++程序设计教程习题答案和实验指导

（第二版）

主 编 瞿绍军 罗 迅 刘 宏

华中科技大学出版社
中国·武汉

内 容 简 介

紧密结合目前高校计算机教学要求和发展趋势,让学生养成良好的编程习惯和编程思维,提高学生分析问题和解决问题的能力,是本书的创新之处。

本书是《C++程序设计教程》(第二版)的配套教材,也可独立使用。全书共分四部分。第一部分为开发环境介绍。第二部分为实验指导,共 13 章,习题均按照 ACM 国际大学生程序设计竞赛标准设计,选用的试题均具有代表性。第三部分为参考答案,其中部分习题提供了多种解决方法。第四部分为五套笔试模拟试卷和三套上机实验考试模拟试卷,并附有参考答案,用来检验学生课程学习的掌握程度,可作为考前的复习,也可供出卷参考。最后为附录。

本课程设有专门的课程学习网站,所有的习题均可在学习网站(http://acm.hunnu.edu.cn/online/? action=course&type=list&coursetype=1)进行测试。

本书特别适合用作计算机专业及其他相关专业的教材,作为 ACM 国际大学生程序设计竞赛入门教材,也可作为各类考试培训和 C++自学教材。

图书在版编目(CIP)数据

C++程序设计教程习题答案和实验指导/瞿绍军,罗迅,刘宏主编. —2 版. —武汉:华中科技大学出版社,2018.4
 ISBN 978-7-5680-3794-5

Ⅰ.①C… Ⅱ.①瞿… ②罗… ③刘… Ⅲ.①C 语言-程序设计-高等学校-教学参考资料
Ⅳ.①TP312.8

中国版本图书馆 CIP 数据核字(2018)第 063593 号

C++程序设计教程习题答案和实验指导(第二版)   瞿绍军 罗迅 刘宏 主编

策划编辑:范　莹
责任编辑:陈元玉
封面设计:原色设计
责任监印:周治超
出版发行:华中科技大学出版社(中国•武汉)   电话:(027)81321913
　　　　　武汉市东湖新技术开发区华工科技园   邮编:430223
录　　排:华中科技大学惠友文印中心
印　　刷:武汉市籍缘印刷厂
开　　本:787mm×1092mm　1/16
印　　张:16.75
字　　数:394 千字
版　　次:2018 年 4 月第 2 版第 1 次印刷
定　　价:39.80 元

本书若有印装质量问题,请向出版社营销中心调换
全国免费服务热线:400-6679-118   竭诚为您服务
版权所有　侵权必究

高等院校计算机系列教材
编 委 会

主　任：刘　宏

副主任：全惠云　熊　江

编　委：（以姓氏笔画为序）

王　毅	王志刚	乐小波	冯先成	刘　琳
刘先锋	刘连浩	羊四清	许又全	阳西述
李　浪	李华贵	李勇帆	杨凤年	肖晓丽
邱建雄	何迎生	何昭青	张　文	张小梅
陈书开	陈倩诒	罗新密	周　昱	胡玉平
徐长梅	徐雨明	高金华	郭广军	唐德权
黄同成	龚德良	符开耀	谭　阳	谭敏生
戴经国	瞿绍军			

前　言

C++语言是目前最流行的面向对象程序设计语言之一。它既支持传统的面向过程的程序设计方法，也支持新的面向对象的程序设计方法。它是 Linux 和 UNIX 下编程的最主要的语言，也是嵌入式开发最常用的编程语言。C++语言全面兼容 C 语言，熟悉 C 语言的程序员仅需学习 C++语言的面向对象特征，就可以很快地使用 C++语言编写程序。

本书是一本通过编程实践引导学生掌握 C++程序设计的教材。我们组织了多位长期从事程序设计、数据结构、面向对象程序设计和计算机算法设计课程教学的老师进行编写，他们都是高校的 ACM 程序设计集训队的教练和指导老师，都有着丰富的教学和编程经验。在写作中，我们力求将复杂的概念用简洁、通俗的语言进行描述，做到深入浅出、循序渐进，争取让学生体会到学习编程的乐趣。

本书将 ACM 国际大学生程序设计竞赛引入课程学习中，使学生从编程入门开始就养成良好的编程习惯和编程思维，强化提升学生分析问题和解决问题的能力，激发学生对编程的兴趣，达到以教学促竞赛、以竞赛强化教学的目的。

ACM 国际大学生程序设计竞赛（简称 ACM-ICPC）是由国际计算机界具有悠久历史的权威性组织 ACM（Association for Computing Machinery，国际计算机学会）主办的，是世界上公认的规模最大、水平最高、参与人数最多的大学生程序设计竞赛，其宗旨是使大学生通过计算机充分展示自己分析问题和解决问题的能力。现在各个高校都非常重视计算机程序设计竞赛。

在平时的教学中，很多同学问怎么才能学好编程？我给他们的答复是"编程再编程"，要想学好一门编程语言，上机动手编写程序是唯一的途径。我们希望你在学习时能把本书的所有习题都自己动手实现并真正掌握，在 OJ 上全部测试通过。

本书是《C++程序设计教程》（第二版）的配套教材，也可独立使用。全书共分为四部分。第一部分为开发环境介绍。第二部分为实验指导，共 13 章，习题均按照 ACM 国际大学生程序设计竞赛标准设计。第 1 章为 C++语言概述，第 2 章为 C++语言编程基础，第 3 章为数组与字符串，第 4 章为函数，第 5 章为指针，第 6 章为结构体与共用体，第 7 章为类与对象及封装性，第 8 章为类的深入，第 9 章为运算符重载，第 10 章为继承性，第 11 章为多态性，第 12 章为输入/输出流，第 13 章为模板和标准库。第三部分为参考答案，其中部分习题提供了多种解决方法。第四部分为五套笔试模拟试卷和三套上机实验考试模拟试卷，并附有参考答案，用来检验学生课程学习的掌握程度，可作为考前的复习，也可供出卷参考。最后为附录，包括 ASCII 码对照表、C/C++与标准 C++头文件对照表、Linux、UNIX 下编译 C++程序、在 Visual C++下调试程序和 Dev-C++调试。

参与本书编写的人员有：瞿绍军、罗迅和刘宏。

本书特别适合用作计算机专业及其他相关专业的教材，可作为 ACM 国际大学生程序设计竞赛入门教材，也可作为各类考试培训和 C++自学教材。

本书的出版得到了湖南师范大学教学改革研究项目"程序设计类课程实践教学体系、内容、方法和手段改革的研究与实践"的资助。

如果你在使用过程中发现错误或有任何疑问，可发邮件给我们反馈和交流(Email:powerhope@163.com)。

编　者

2018 年 2 月

目 录

第一部分 开发环境 ... 1

C++程序的开发环境 ... 2
- 一、Microsoft Visual C++ ... 2
- 二、Microsoft Visual Studio 2010 ... 6
- 三、Dev-C++ ... 12
- 四、CodeBlocks ... 15
- 五、在线评测系统使用 ... 19

第二部分 实验指导 ... 25
- 第 1 章 C++语言概述 ... 26
- 第 2 章 C++语言编程基础 ... 26
- 第 3 章 数组与字符串 ... 31
- 第 4 章 函数 ... 39
- 第 5 章 指针 ... 44
- 第 6 章 结构体与共用体 ... 51
- 第 7 章 类与对象及封装性 ... 53
- 第 8 章 类的深入 ... 55
- 第 9 章 运算符重载 ... 58
- 第 10 章 继承性 ... 60
- 第 11 章 多态性 ... 64
- 第 12 章 输入/输出流 ... 67
- 第 13 章 模板和标准库 ... 67

第三部分 参考答案 ... 75
- 第 1 章 C++语言概述 ... 76
- 第 2 章 C++语言编程基础 ... 77
- 第 3 章 数组与字符串 ... 87
- 第 4 章 函数 ... 100
- 第 5 章 指针 ... 111
- 第 6 章 结构体与共用体 ... 125
- 第 7 章 类与对象及封装性 ... 131
- 第 8 章 类的深入 ... 135
- 第 9 章 运算符重载 ... 142
- 第 10 章 继承性 ... 155

第 11 章 多态性 161
第 12 章 输入/输出流 166
第 13 章 模板和标准库 169

第四部分 模拟试卷 181

笔试模拟试卷（1） 182
笔试模拟试卷（2） 191
笔试模拟试卷（3） 200
笔试模拟试卷（4） 205
笔试模拟试卷（5） 210
上机实验考试模拟试卷（1） 215
上机实验考试模拟试卷（2） 216
上机实验考试模拟试卷（3） 218
笔试模拟试卷（1）参考答案 220
笔试模拟试卷（2）参考答案 225
笔试模拟试卷（3）参考答案 230
笔试模拟试卷（4）参考答案 231
笔试模拟试卷（5）参考答案 232
上机实验考试模拟试卷（1）参考答案 233
上机实验考试模拟试卷（2）参考答案 236
上机实验考试模拟试卷（3）参考答案 240

附 录 245

附录 A ASCII 码对照表 246
附录 B C/C++与标准 C++头文件对照表 247
附录 C Linux、UNIX 下编译 C++程序 248
附录 D 在 Visual C++下调试程序 252
附录 E Dev-C++调试 257

第一部分 开发环境

C++程序的开发环境

支持 C++程序开发的工具很多，比较流行的 C++程序集成开发环境有基于 Window 平台的 Microsoft Visual C++、Microsoft Visual Studio 系列、CodeBlocks 和 Dev-C++等；基于 Window 平台、Linux 及 UNIX 下的 Eclipse、CodeBlocks 和 NetBeans 等 IDE。下面分别对 Microsoft Visual C++、CodeBlocks 和 Dev-C++开发环境的使用进行简要介绍，读者可以根据自己的爱好选择对应的开发环境。如果学习后准备参加大学生程序设计竞赛，则应该熟练使用 CodeBlocks。

一、Microsoft Visual C++

Microsoft Visual C++是美国 Microsoft 公司最新推出的可视化 C++开发工具，是目前计算机开发者首选的 C++开发环境。它支持最新的 C++标准，它的可视化工具和开发向导使 C++应用开发变得非常方便、快捷。

Microsoft Visual C++已经从 Visual C++ 1.0 版本发展到最新的 Visual Studio 2015 版本。本节将以 Visual C++ 6.0 和 Visual Studio 2010 为背景介绍 Visual C++的使用方法，Visual Studio 2010 后续版本的使用与 Visual Studio 2010 的差别不大。

1．启动 Microsoft Visual C++

当 Microsoft Visual C++成功安装后，在 Windows 桌面依次选择"开始"→"所有程序"→"Microsoft Visual Studio 6.0"→"Microsoft Visual C++ 6.0"，可以启动 Microsoft Visual C++ 6.0。Microsoft Visual C++ 6.0 的集成开发环境如图 1.1 所示。

2．创建工程

在 Microsoft Visual C++环境中，开发应用程序的第一步是创建一个工程。Microsoft Visual C++使用工程组织和维护应用程序。工程文件保存了与工程有关的信息。每个工程都保存在自己的目录中。每个工程目录包括一个工作区文件（.dsw）、一个工程文件（.dsp）、至少一个 C++程序文件（.cpp）以及 C++头文件（.h）。

（1）依次单击"File"→"New"，如图 1.2 所示。

（2）在弹出的对话框中单击"Projects"选项卡，选中"Win32 Application"，在"Project name"中输入工程名，然后在"Location"中选择工程保存的位置。最后单击"OK"按钮，如图 1.3 所示。

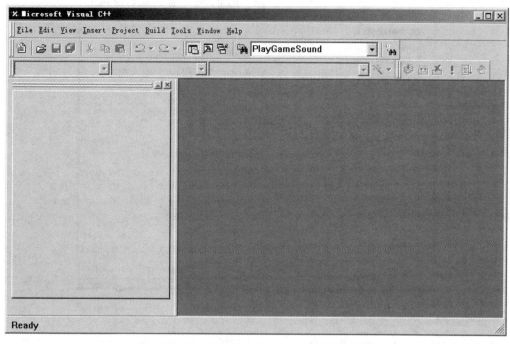

图 1.1 Microsoft Visual C++ 6.0 集成开发环境

图 1.2 "File"菜单

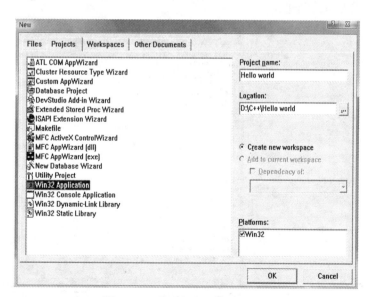

图 1.3 Visual C++ 6.0 向导

（3）此时出现如图 1.4 所示的"Win32 Application - Step 1 of 1"对话框，选择"An empty project."单选项，单击"Finish"按钮。

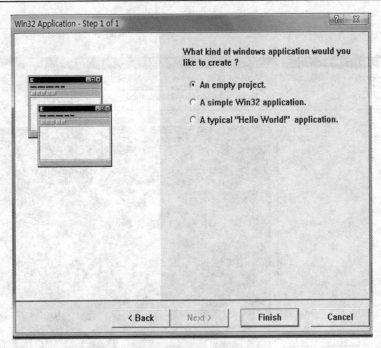

图 1.4　控制台工程向导

（4）出现如图 1.5 所示的"New Project Information"对话框，单击"OK"按钮，完成工程的创建。

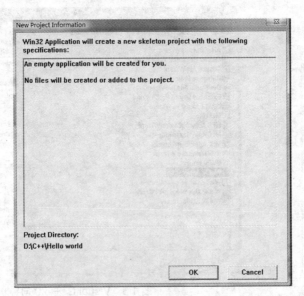

图 1.5　工程信息

3. 编辑 C++源程序

（1）单击"File"→"New"，在弹出的对话框中单击"Files"选项卡，选中"C++ Source

File",选中"Add to project",在"File"文本框中输入文件名,然后在"Location"文本框中选择文件保存的位置(用默认即可,和工程保存在同一位置)。最后单击"OK"按钮,如图 1.6 所示。

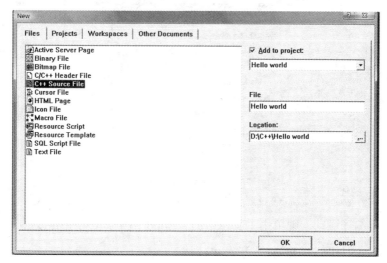

图 1.6　新建 C++源程序文件

(2)在编辑区输入如图 1.7 所示的代码。输入完毕后,再单击"File"菜单下的"Save"子菜单,保存代码。

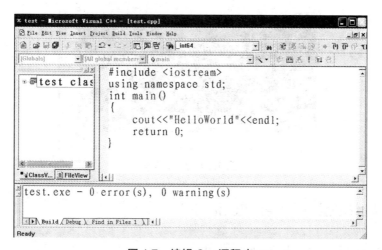

图 1.7　编辑 C++源程序

4．编译和运行

(1)单击工具栏中的"Compile"图标(见图 1.8)或选择"Build"菜单下的"Compile Hello world.cpp"子菜单,或按 Ctrl+F7 快捷键,如图 1.9 所示。

图 1.8　工具栏

图 1.9　"Build"菜单

（2）如果编译成功，则单击工具栏中的"Build"图标（见图 1.8）或选择"Build"菜单下的"Build Hello world.exe"子菜单或按 F7 键。

（3）单击工具栏中的"Execute program"图标（见图 1.8）或选择"Build"菜单下的"Execute Hello world. exe"子菜单或按 Ctrl+F5 快捷键。

（4）运行结果如图 1.10 所示。

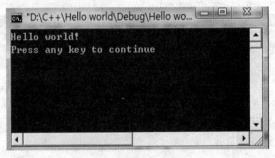

图 1.10　运行结果

二、Microsoft Visual Studio 2010

1. 启动 Microsoft Visual Studio 2010

当 Microsoft Visual Studio 2010 成功安装后，在 Windows 桌面依次选择"开始"→"所有程序"→"Microsoft Visual Studio 2010"，可以启动 Microsoft Visual Studio 2010。Microsoft Visual Studio 2010 的集成开发环境如图 1.11 所示。

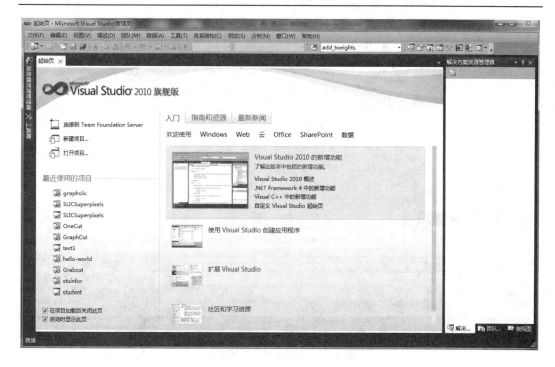

图 1.11 Microsoft Visual Studio 2010 的集成开发环境

2. 创建项目

(1) 依次单击"文件"→"新建"→"项目",如图 1.12 所示。

(2) 弹出的"新建项目"对话框如图 1.13 所示,在左侧"已安装的模板"中选择"Visual C++"→"Win32",右侧选择"Win32 控制台应用程序",在下面的"名称"栏后输入项目名称,"位置"栏中选择项目的保存位置,或单击"浏览(B)…"按钮进行选择,然后选中"为解决方案创建目录(D)",最后单击"确定"按钮。

(3) 出现如图 1.14 所示的"Win32 应用程序向导-HelloWorld"对话框,单击"下一步"按钮。

图 1.12 "新建"菜单

图 1.13 "新建项目"对话框

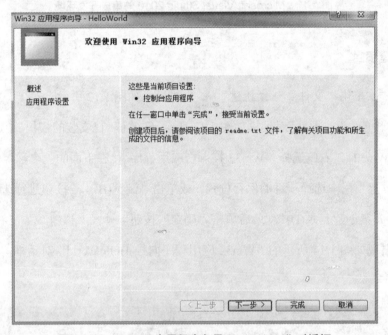

图 1.14 "Win32 应用程序向导-HelloWorld"对话框

（4）在如图 1.15 所示的"Win32 应用程序向导-HelloWorld"对话框的"应用程序类型"下选择"控制台应用程序"，在"附加选项"下选择"空项目"。最后单击"完成"按钮完成项目的创建工作。

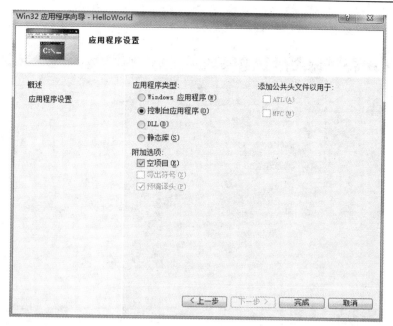

图 1.15 "Win32 应用程序向导-HelloWorld"项目创建完成

3. 编辑 C++源程序

（1）在窗口右边"解决方案资源管理器"中右键单击"源文件"，依次选择"添加"→"新建项(W)…"，如图 1.16 所示，或单击"项目"菜单下的子菜单"添加新项"。

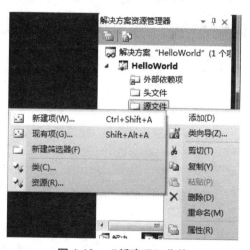

图 1.16 "新建项"菜单

（2）弹出的"添加新项-HelloWorld"对话框如图 1.17 所示，在左侧的"已安装的模板"下选择"Visual C++"→"代码"，在右侧中选择"C++文件(.cpp)"，下面的"名称"栏后输入文件名称，"位置"栏后选择保存路径（建议用默认，和项目在同一目录下),最后单击"添加"按钮。

图 1.17 "添加新项-HelloWorld"对话框

(3) 在代码编辑区输入如图 1.18 所示的代码。再单击"文件"菜单下的"保存"子菜单,保存好源代码。

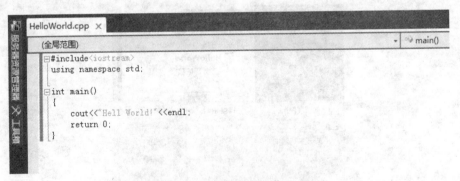

图 1.18 编辑 C++源代码

4. 编辑和运行

(1) 在"生成"菜单中单击"生成解决方案"或按 F7 键,如图 1.19 所示。

(2) 编译成功后,单击"调试"菜单中的"开始执行(不调试)",或按 Ctrl+F5 快捷键执行程序,如图 1.20 所示。

(3) 运行结果如图 1.21 所示。

图 1.19 "生成"菜单

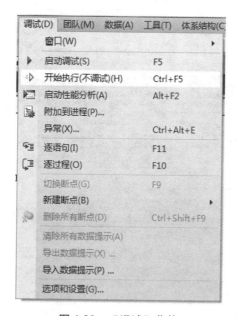

图 1.20 "调试"菜单

图 1.21 运行结果

提示:从 Visual Studio 2005 开始,要编译和运行 C++程序,必须先创建项目,否则无法编译和运行。

三、Dev-C++

Dev-C++是一个 Windows 下的 C 和 C++程序的集成开发环境。它使用 MingW32/GCC 编译器,遵循 C/C++标准。开发环境包括多页面窗口、工程编辑器以及调试器等,在工程编辑器中集合了编辑器、编译器、连接程序和执行程序,提供了高亮度的语法显示,以减少编辑错误,还提供了完善的调试功能,可适合初学者与编程高手的不同需求,是学习 C 或 C++的首选开发工具。

Dev-C++还是很多程序设计竞赛提供的比赛环境。

1. 启动 Dev-C++

当 Dev-C++成功安装后,通过选择 Windows 桌面的"开始"→"所有程序"→"Bloodshed Dev-C++"→"Dev-C++",可以启动 Dev-C++。Dev-C++的集成开发环境如图 1.22 所示。

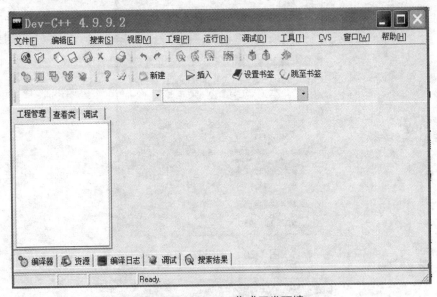

图 1.22 Dev-C++集成开发环境

2. 创建工程

在 Dev-C++环境中,开发应用程序的第一步是创建一个工程。

(1)依次单击"文件"→"新建"→"工程",选择"Console Application",在"名称"文本框中输入工程的名称,选择"C++工程"选项,如图 1.23 所示。

(2)单击"确定"按钮,系统就会创建好工程,并自动创建一个"main.cpp"的源程序文件,如图 1.24 所示。

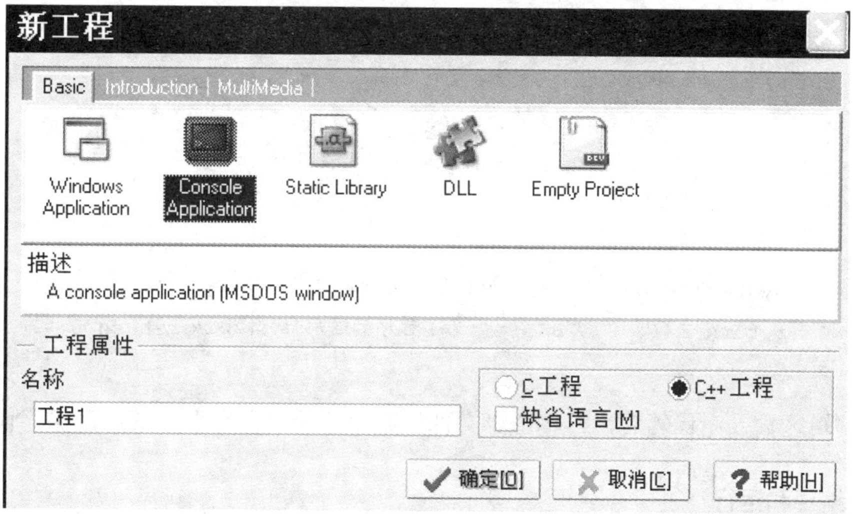

第一部分　开发环境

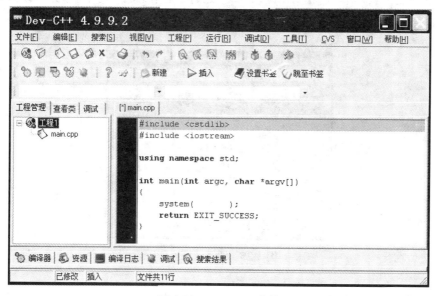

图 1.23　"新工程"对话框

图 1.24　main.cpp 文件

这里我们也可以不创建工程，而直接创建一个 C++源程序文件，依次单击"文件"→"新建"→"源代码"，即可创建一个空的源程序文件，然后自己在里面输入源程序。

3. 编辑 C++源程序

在如图 1.24 所示的编辑区中输入如下代码：

```
#include <cstdlib>    //后面用到的 system 函数所在的库
#include <iostream>   //输入/输出流
```

```
using namespace std;  //命名空间

int main(int argc, char *argv[])    //主函数
{
    cout<<"HelloWorld"<<endl;
    system("PAUSE");  //避免在程序运行时一闪而过
    return EXIT_SUCCESS;  //为了程序的通用性，修改成 return 0;
}
```

在编辑区输入完代码后，再单击"文件"菜单下的"保存"子菜单，保存代码。

4．编译和运行

（1）单击工具栏中的"编译"图标，或"运行"菜单下的"编译"子菜单或按 **Ctrl+F9** 快捷键编译程序，如果编译成功，则显示结果如图 1.25 所示。

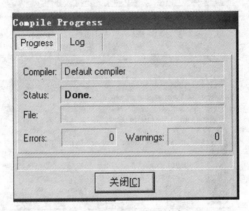

图 1.25　编译结果信息

（2）运行程序，单击工具栏中的"运行"图标或"运行"菜单下的"运行"子菜单或按 **Ctrl+F10** 快捷键，运行程序。

（3）运行结果如图 1.26 所示。

图 1.26　运行结果

四、CodeBlocks

Code::Blocks 可简写为 CodeBlocks，是一个开源的 C++集成开发环境。CodeBlocks 可跨平台支持，不仅支持 Linux 和 Windows 系统，也支持 Mac 系统。对于 C++语言，CodeBlocks 包括对 Borland C++、VC++、Inter C++等 20 多个不同厂家或版本编译器的支持。另外，CodeBlocks 也支持多种编程语言的编译，包括"D"语言。通常情况下，我们采用开源 g++ 编译器作为 C++默认的编译器。在 Linux 下，g++由操作系统自带。Windows 环境下，需要 mingw32 库支持。但 Code::Blocks 在安装包中已经自带了 mingw32 的库文件。

1. 启动 CodeBlocks

当 CodeBlocks 成功安装后，如在 Windows 7 中，通过选择 Windows 桌面的"开始"→"所有程序"→"CodeBlocks"→"CodeBlocks"，启动 CodeBlocks。它的集成开发环境如图 1.27 所示。

图 1.27　CodeBlocks 集成开发环境

2. 创建工程和源文件

在 CodeBlocks 环境中，开发应用程序的第一步是创建一个工程。

（1）依次单击"File"→"New"→"Project"，如图 1.28 所示。

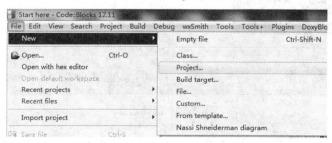

图 1.28　"New"菜单

选择"Console application",如图 1.29 所示,然后单击"Go"按钮,再次出现"Console application"信息对话框,直接单击"Next"按钮。

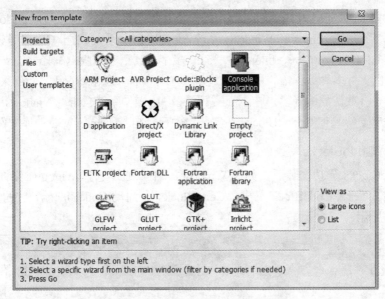

图 1.29　"New from template"对话框

(2) 在"Console application"下选择"C++",再单击"Next"按钮,如图 1.30 所示。

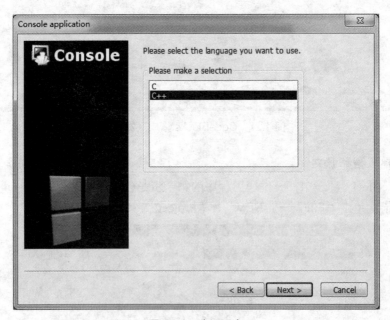

图 1.30　选择语言

(3) 在"Console application"对话框的"Project title"文本框中输入工程的名称,

"Folder to create project in:"下拉框中选择工程的存放位置,下面会自动生成工程文件名 *.cbp(cbp 是 Code::Blocks Project),再单击"Next"按钮,如图 1.31 所示。

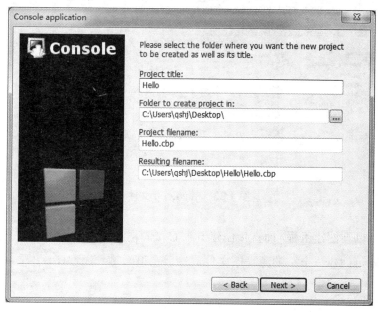

图 1.31 工程名称和位置设置对话框

(4)选择编译器,一般选择"GNU GCC Compiler",其他采用默认即可,再单击"Finish"按钮,如图 1.32 所示。

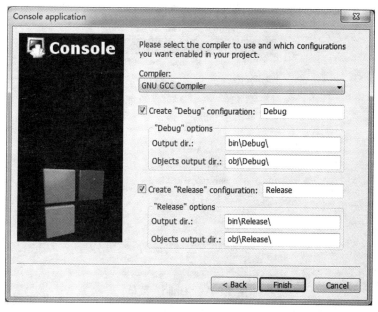

图 1.32 选择编译器

（5）系统创建好工程，并自动创建一个"main.cpp"源程序文件，如图1.33所示。

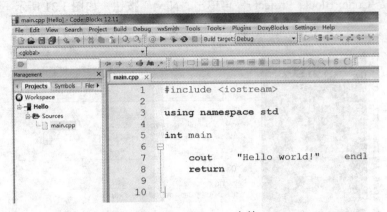

图 1.33 main.cpp 文件

这里也可以不创建工程，而直接创建一个C++源程序文件，依次单击"File"→"New"→"File"，即可创建一个空的源程序文件，然后在里面输入源程序。

2．编译和运行

（1）单击工具栏中的"Build"图标，或"Build"菜单下的"Build"子菜单，或按Ctrl+F9快捷键编译程序，如图1.34所示，如果编译成功，则显示如图1.35所示的结果，如果编译出错，则会出现错误提示信息。

图 1.34 "Build"菜单

（2）运行程序，单击工具栏中的"Run"图标，或"Build"菜单下的"Run"子菜单，或按 Ctrl+F10 快捷键，运行程序，如图 1.36 所示。

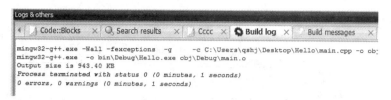

图 1.35　编译结果信息

（3）运行结果如图 1.36 示。

图 1.36　运行结果

有关程序的调式请查看附录 B 和附录 C。

五、在线评测系统使用

本书的所有习题都可以在湖南师范大学在线评测系统上提交测试。下面以湖南师范大学在线评测系统为例介绍。网址为：http://acm.hunnu.edu.cn/。进入网页后，单击网站导航条右边的"Judge Online"进入在线评测系统，如图 1.37 所示。

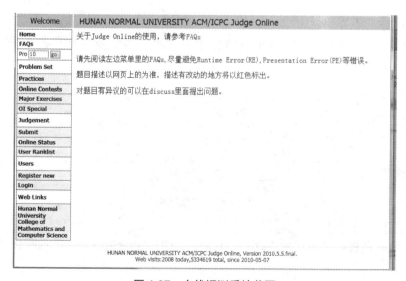

图 1.37　在线评测系统首页

（1）注册。单击"Register new"，打开"用户注册网页"。注册信息时有"*"的内容必须填写，如图1.38所示。

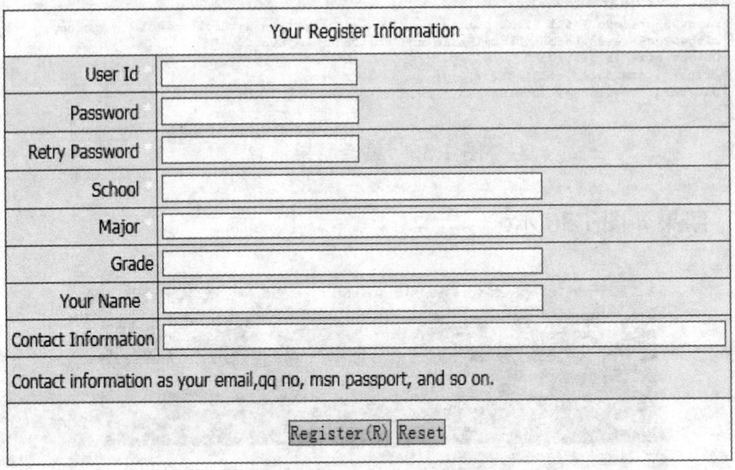

图1.38 用户注册网页

填写好后单击"Register"按钮进行注册，注册成功后会出现"Register OK"提示信息，如图1.39所示。

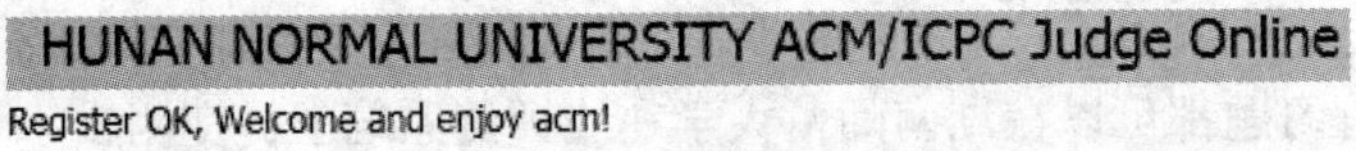

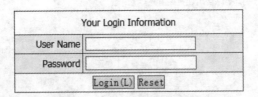

图1.39 注册成功信息

（2）登录。单击网页左侧的"Login"，输入自己注册时的用户名（User Name）和密码（Password），再单击"Login"按钮，如图1.40所示。如果用户名和密码都正确，将会成功登录。

图1.40 登录网页

（3）查看题目。单击左侧"Problem Set"下的"Practices"，进入题目列表，如图1.41所示。

然后点题号 10000 或"Title"下的"An Easy Problem",进入对应的题目,如图 1.42 所示。

Solved	Problem	AC/Submit	Title	Source
	10000	1063/1336	An Easy Problem	HNU Contest
	10001	284/352	阅读顺序	HNU Contest
	10004	40/46	Array	HNU Contest
	10005	14/24	The milliard Vasya's function	HNU Contest
	10006	13/16	Computer	HNU Contest
	10007	1/5	Robot In The Field	HNU Contest
	10008	22/52	Equilateral triangle	HNU Contest
	10009	1/2	Fragment Assembly	HNU Contest
	10010	4/6	Towards Zero	HNU Contest
	10011	47/67	Hamming Distance	HNU Contest
	10012	24/26	Overlapping Rectangles	HNU Contest
	10013	1/6	James Bond	HNU Contest
	10014	0/2	检查次品	DR.Wu
	10015	116/144	大数的乘法	qshj
	10016	32/33	Beautiful Meadow	ZOJ Monthly, June 2009
	10017	61/76	谁拿了最多奖学金	NOIP 2005
	10019	27/32	Calendar	Shanghai-P 2004
	10020	40/40	Self Numbers	MCU1998
	10021	66/69	Lowest Bit	HNU 1'st Contest
	10022	26/28	Magician	HNU 1'st Contest
	10023	9/10	SpinLock	HNU 1'st Contest

图 1.41　题目列表

An Easy Problem

Time Limit: 1000ms, **Special Time Limit:** 2500ms, **Memory Limit:** 32768KB
Total submit users: 1336, **Accepted users:** 1063
Problem 10000 : No special judgement

Problem description
We just test the input and output system in your selected language.

Input
The Input will contain only two integers a,b.

Output
The Ouput should contains the result of a + b. No other whitespace contians at begin or end of the output Notice : empty line is allowed.

Sample Input
10 20

Sample Output
30

Problem Source
HNU Contest

Submit　Discuss　Judge Status　Problems　Ranklist

图 1.42　查看题目

图 1.42 所示题目的解答可查看左边的 FAQ,对应不同语言的具体规范在里面有详细的说明,如图 1.43 所示。

(4)提交。在问题提交前,一般先到自己的机器上写好程序,再用对应编译器进行编译,如果编译正确,接着测试"Sample Input"里面的输入数据,如果输出结果和"Sample

Output"一致,再进行提交。在图 1.42 中单击"Submit"按钮,会出现如图 1.44 所示的网页,然后把你写好的程序拷贝到下面的文本框中。

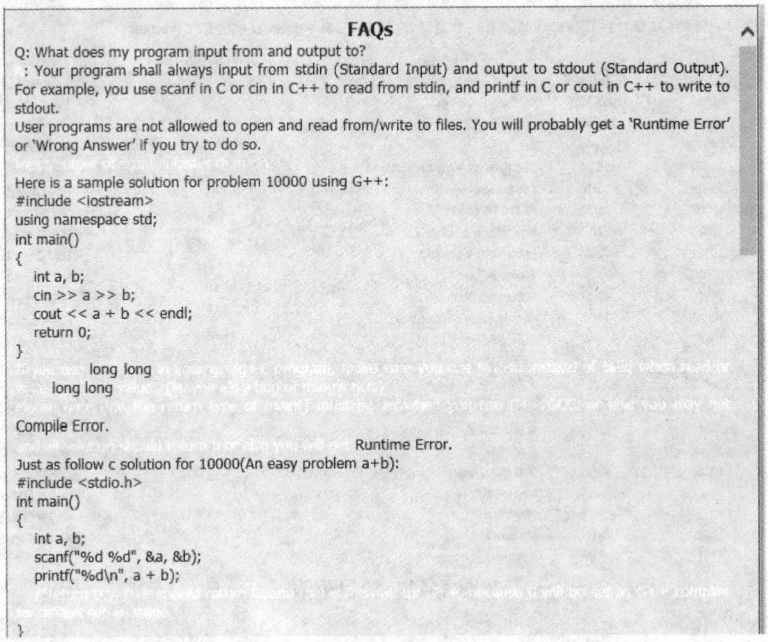

图 1.43 FAQs

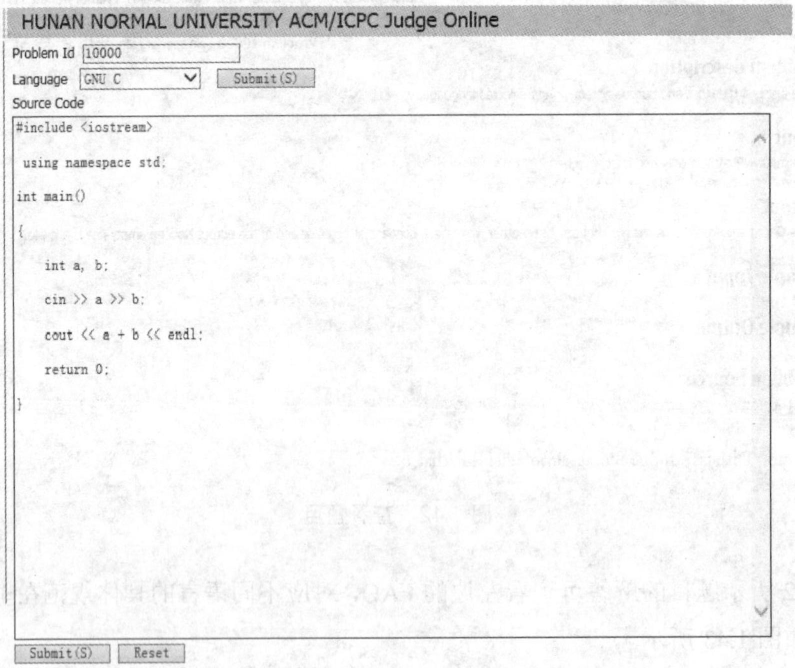

图 1.44 源代码编辑框

提交前需要在“Language”选择框中选择你所采用的编程语言，默认选择的是“GNU C”，这里请点击向下箭头，在下拉选择框中选择“GNU C++”，然后点击“Submit(s)”按钮，会出现如图 1.45 所示的评估状态网页。

HUNAN NORMAL UNIVERSITY ACM/ICPC Judge Online

Realtime judge status

Solution	User	Problem	Language	Judge Result	Memory	Time Used	Code Length	Submit Time
106478	admin	10000	GNU C++	Waiting	0KB	0ms	156B	2014-11-28 21:53:30.0

图 1.45 Realtime judge status

刚提交时反馈给你的状态“Judge Result”为“Waiting”，表示服务器正在判断你提交的程序，几秒钟后，刷新一下你的网页（或按 F5 键），状态会发生改变，如果反馈的结果为“Accepted”，则表示你的程序通过了服务器上的所有测试，结果为正确。

第二部分　实验指导

习题和湖南师范大学 OJ 上对应的题号对照表如下。

习题	OJ 对应题号	习题	OJ 对应题号	习题	OJ 对应题号
1.5	10000				
2.1	10452	2.2	10454	2.3	10459
2.4	10461	2.5	10464	2.6	11601
2.7	11602	2.8	10453	2.9	10454
2.10	11603	2.11	10458	2.12	11604
2.13	11605	2.14	11606		
3.1	10477	3.2	10475	3.3	10474
3.4	10472	3.5	10471	3.6	10468
3.7	10469	3.8	10473	3.9	10476
3.10	10478	3.11	10484	3.12	10487
3.13	10367	3.14	11607		
4.1	10452	4.2	10454	4.3	10459
4.4	10461	4.5	10464	4.6	11601
4.7	11602	4.8	10453	4.9	10454
4.10	11603	4.11	10458	4.12	11604
4.13	11605				
5.1	10479	5.2	10480	5.3	10481
5.4	10482	5.5	10483	5.6	10484
5.7	10485	5.8	10486	5.10	10487
6.1	10488	6.2	10489	6.3	10490
6.4	10491	6.5	10492		
7.2	10495	7.3	10496		
8.2	10497	8.3	10498	8.4	11608
8.5	11609	8.6	11610		
9.1	10499	9.2	10500	9.3	10502
10.2	10503	10.3	11611	10.4	11612
11.1	11613	11.2	11614	11.3	11615
11.4	11616				
13.1	10543	13.2	10595	13.3	10895
13.4	10898	13.5	10919	13.6	11068
13.7	11208	13.8	11294	13.9	11298
13.10	11617				

第 1 章 C++语言概述

1.1 熟悉 C++语言的集成开发环境。
1.2 掌握 C++应用程序的编辑、编译、链接和运行过程。
1.3 将本章中的例子程序在 C++应用环境中进行测试。
1.4 熟悉 OJ 的使用，在 OJ 上注册自己的账号。
1.5 在湖南师范大学 OJ 上完成题号为"10000"的题目。

第 2 章 C++语言编程基础

2.1 题号：10452

题目描述：求三个数中最大数的数值。

输入：输入有若干行，每行有三个 32 位整数。

输出：对于每一行输入，输出一行，输出结果为该行输入的三个数中最大数的数值。

样例输入：

```
2 5 9
8 24 1
```

样例输出：

```
9
24
```

2.2 题号：10454

题目描述：求 n 的阶乘。

输入：输入的第一行是一个正整数 m。接下来有 m 行，每行一个整数 n，$1 \leq n \leq 10$。

输出：对于每个 n 输出一行，输出结果为 n!。

样例输入：

```
3
1
5
10
```

样例输出：

```
1
```

```
120
3628800
```

2.3 题号：10459

题目描述：本题中的水仙花数是指一个三位正整数，其各位数字的立方和等于该数。例如：407=4*4*4+0*0*0+7*7*7，所以 407 是一个水仙花数。判断给定的数 n 是否为水仙花数。

输入：输入有若干行，每行一个整数 n，$1 \leq n \leq 999$。

输出：对于每个输入，输出一行，如果是水仙花数，则输出"YES"，否则输出"NO"。

样例输入：
```
153
100
```

样例输出：
```
YES
NO
```

2.4 题号：10461

题目描述：完数是指不包括其本身的所有因子之和恰好等于其本身的数。例如 6 就是一个完数，因为 6 的所有因子（除 6 本身外）是 1、2、3，其和恰好是 6。给定整数 n，判断是否为完数。

输入：输入有若干行，每行一个正的 32 位整数 n。

输出：对每个输入，输出一行。首先输出"Case 序号:"，然后输出 n 和逗号，再输出"Yes"或"No"。

样例输入：
```
6
9
15
```

样例输出：
```
Case 1: 6,Yes
Case 2: 9,No
Case 3: 15,No
```

2.5 题号：10464

题目描述：素数是只能被 1 和本身整除的整数。例如 2、3、5、7 是素数，4、6、8、9 不是素数。判断给定的整数 n 是否为素数。

输入：第一行是一个正整数 T，其后有 T 行，每行一个正整数 n，n<10000。

输出：对于每个输入 n，输出一行，若 n 是素数，则输出 1，否则输出 0。

样例输入：

```
    3
    2
    7
    9
```
样例输出：
```
    1
    1
    0
```

2.6 题号：11601

题目描述：求 n 个数的平均值。

输入：输入有若干行，每行第一个是一个正整数 n（n<1000），其后还有 n 个 32 位整数。

输出：对于每一行输入，输出这 n 个数的平均值，保留两位小数。

样例输入：
```
    2 3 4
```
样例输出：
```
    3.50
```

2.7 题号：11602

题目描述：有 n 个数，求其最小值。

输入：输入有若干行，每行第一个数是一个正整数 n（n<1000），其后还有 n 个 32 位整数。

输出：对于每一行输入，输出这 n 个数中最小数的数值。

样例输入：
```
    2 3 4
```
样例输出：
```
    3
```

2.8 题号：10453

题目描述：有如下分段函数。给定自变量，计算因变量的值。

$$y = \begin{cases} x & (x<1) \\ 2x-1 & (1 \leqslant x<10) \\ 3x-11 & (x \geqslant 10) \end{cases}$$

输入：输入第一行是一个正整数 n，其后有 n 行。每行一个数为 x。

输出：根据 x 计算 y 值。如果 y 值是整数，则直接输出；否则保留一位小数，四舍五入。

样例输入：

```
   2
   5
  30
```
样例输出：
```
   9
  79
```

2.9 题号：10454

题目描述：给定如下表达式及其参数，计算表达式的值。

$$\frac{(x+1)(y-3)}{x+y+z}$$

输入：输入第一行为一个正整数 n，其后有 n 行。每行三个数，分别是 x、y、z。

输出：对每一组 x、y、z，输出一行为表达式的值。保留四位小数。

样例输入：
```
   2
   1 2 3
   7 8 20
```
样例输出：
```
  -0.3333
   1.1429
```

2.10 题号：11603

题目描述：判断闰年。

输入：输入有若干行，每行一个正整数 y，且 y≤9999。

输出：对每一个输入，输出一行。y 是闰年则输出"Yes"，否则输出"No"。

样例输入：
```
  1996
  1997
```
样例输出：
```
  Yes
  No
```

2.11 题号：10458（因为学生对二进制表示可能不太熟练，所以本题描述较长）

题目描述：计算机中的每一个数都是用二进制表示的。如数值 5，将其看做是一个 32 位整型数，在计算机中就可表示为：0000 0000 0000 0000 0000 0000 0000 0101。一共是 32 个二进制位。

现在将其循环右移一位，将变成：1000 0000 0000 0000 0000 0000 0000 0010。这个数是多少呢？如果是 int 类型，这个数是 -2147483646。

如果将 5 循环右移 2 位，将变成：0100 0000 0000 0000 0000 0000 0000 0001。这个数就变成 1073741825。

非常奇妙，对 int 做循环移位，将使得结果在正负之间来回变换。循环左移的原理是一样的。而且，你应该能够发现把 5 循环左移 31 位，就等于将其循环右移 1 位。当然，这也与 5 用 32 位表示有关，如果用 64 位表示就不是这样了。

输入：输入第一行为一个正整数 T，其后有 T 行。每行两个 32 位整数 n、m。n≥0 时，表示将 m 循环右移 n 位；否则，将 m 循环左移 n 位。

输出：对每个输入，输出一行，为移位后的结果。

样例输入：

2
1 2
-2 2

样例输出：

1
8

2.12　题号：11604

题目描述：很多公式可以用来计算 π，甚至还可以用概率的方法。这里提供一个最简单的公式：

$$\pi/4 = 1 - 1/3 + 1/5 - 1/7 + 1/9 - \cdots$$

请利用该公式将 π 计算到指定位数。

输入：输入有若干行，每行一个非负整数 n，表示输出 π 的小数位数，n 小于等于 12。n 等于 -1 时，测试结束。

输出：对每个输入（-1 除外），输出对应小数位数的 π，四舍五入。

样例输入：

2
3
-1

样例输出：

3.14
3.142

2.13　题号：11605

题目描述：计算完 π，来计算一下自然对数的底 e 吧。这里有一个简单的公式：

$$e = 1/0! + 1/1! + 1/2! + 1/3! + \cdots$$

输入：输入有若干行，每行一个非负整数 n，表示输出 e 的小数位数，n 小于等于 12。n 等于 -1 时，测试结束。

输出：对每个输入（-1 除外），输出对应小数位数的 e，四舍五入。

样例输入：

    2
    3
    -1

样例输出：

    2.72
    2.718

2.14 题号：11606

题目描述：编写一个可以计算表达式的程序。

输入：输入的数据仅由操作数 a、运算符 c 和操作数 b 三部分组成。其中 a、b 为(0,2^15)之内的正整数，操作符 c 为 "+"、"-"、"*"、"/" 或 "%" 五个字符之一，分别表示加、减、乘、整除、取余。

输出：按照格式输出原表达式及其值。

样例输入：

    2 + 3
    5 / 2
    5 % 2
    4 * 4

样例输出：

    2+3=5
    5/2=2
    5%2=1
    4*4=16

第 3 章 数组与字符串

3.1 题号：10477

题目描述：给定一个非负整数 n，打印杨辉三角前 n 行。

输入：输入数据有若干行。每一行有一个非负整数 n（1≤n≤20）对应一种情形。

输出：对于每一种情形，先输出 "Case #:"（#为序号，从 1 开始），换行；然后输出结果（参见样例输出）。设置 setw(6)使数据占 6 个字符宽；每种情形中，最后一行第一个数字从第 6 列开始，每行最后一个数字后面不需要输出空格。

样例输入：

    3
    4

样例输出：
    Case 1:
            1
          1   1
        1   2   1
    Case 2:
            1
          1   1
        1   2   1
      1   3   3   1

3.2 题号：10475

题目描述：下面是一个 5×5 的螺旋方阵。你的任务是输出逆时针方向旋转的 n×n 螺旋方阵。

    1 16 15 14 13
    2 17 24 23 12
    3 18 25 22 11
    4 19 20 21 10
    5 6 7 8 9

输入：输入文件只有一行，它是由若干个整数 n 组成的（少于 50 个数），且每两个整数之间有一个空格，尾部无多余空格，（1≤n≤70）。

输出：对输入文件中的每个整数 n，先在一行上输出 "n="，再输出 n 的值。接着在下面的 n 行上按 n 行 n 列的方式输出 n×n 螺旋方阵，行尾无空格，同一行上两个数之间空一格。两个螺旋方阵之间空一行。

样例输入：
    4 5
样例输出：
    n=4
    1 12 11 10
    2 13 16 9
    3 14 15 8
    4 5 6 7

    n=5
    1 16 15 14 13
    2 17 24 23 12
    3 18 25 22 11
    4 19 20 21 10

5 6 7 8 9

3.3 题号：10474

题目描述：有 n 个整数，已按从小到大的顺序排列好，再输入一个数，把它插入到原有的数列中，而且仍保持有序，同时输出新数列。

输入：第一行是一个整数 T，表示有 T 组数据。每组数据占两行，第一行：第一个数为 n，表示数列中数的个数，后面是数列中的 n 个数。第二行，待插入有序数列中的一个数。

输出：对于每组测试数据，输出新的数列。数据间用一个空格分隔，行最后一个数据后面无空格，换行。

样例输入：

2
6 1 3 5 8 9 10
2
5 2 10 98 456 871
900

样例输出：

1 2 3 5 8 9 10
2 10 98 456 871 900

3.4 题号：10472

题目描述：已知两个矩阵 **A**、**B**，求 **A** 与 **B** 的乘积矩阵 **C** 并输出结果，其中 **C** 中的每个元素 c[i][j]等于$\sum$**A[i][k]*B[k][j]**。

输入：

第一行是测试数据的组数 T（1≤T≤10）。接着是 T 组测试数据的描述。

每一组数据的第一行是三个整数 p、q、r（p、q、r≤10），表示第一个矩阵的阶为 p×q，第二个矩阵的阶为 q×r。接着的 p 行，每行有 q 个整数，表示第一个矩阵；再空一行。然后的 q 行，每行有 r 个整数，表示第二个矩阵。两组测试数据之间空一行。

输出：对于每组测试数据，先输出"Case #:"（#为序号，从 1 开始），换行，输出矩阵。数据间用一个空格分隔，每行最后一个数据后面无空格。

样例输入：

2
2 2 3
3 4
2 5

3 2 5
5 7 2

```
3 4 3
-4 17 -4 17
7 5 14 22
-7 3 -10 11

-12 3 2
0 14 -3
8 10 -1
8 5 2
```
样例输出：
```
Case 1:
29 34 23
31 39 20
Case 2:
152 271 -21
204 341 29
92 -24 9
```

3.5 题号：10471

题目描述：已知一个数值矩阵，求出该矩阵的转置矩阵并输出结果，其中转置矩阵中的[i][j]位置上的元素等于原矩阵中的[j][i]位置上的元素。

输入：第一行一个整数 T，表示有 T 组数据。每组数据的第一行为两个整数 m、n（m≤100，n≤100），表示矩阵为 m 行 n 列，接下来的 m 行，每行 n 个整数，表示矩阵中的数据，数据间用空格隔开。

输出：对于每组测试数据，先输出"Case #:"（#为序号，从 1 开始），换行，输出矩阵。数据间用一个空格分隔，每行最后一个数据后面无空格。

样例输入：
```
2
2 3
1 2 3
3 4 5
3 3
1 2 3
4 5 6
7 8 9
```
样例输出：
```
Case 1:
```

```
1 3
2 4
3 5
Case 2:
1 4 7
2 5 8
3 6 9
```

3.6 题号：10468

题目描述：有一个数列，它的第一项为 0，第二项为 1，以后每一项都是它的前两项之和，试产生出此数列的前 n 项，并按逆序输出。

输入：多行数据，每行一个整数 n（n≤40），表示数列前 n 项。

输出：按逆序输出此数列前 n 项，数据间用一个空格分隔，最后一个数据后面无空格。

样例输入：
```
5
```
样例输出：
```
3 2 1 1 0
```

3.7 题号：10469

题目描述：输入一个字符串，假定该字符串的长度不超过 256，试统计出该串中所有十进制数字字符的个数。

输入：第一行一个整数 T，表示有 T 组数据。以下 T 行，每行是一个字符串。

输出：对于每组测试数据，输出结果，换行。

样例输入：
```
2
a123
qethghd23fg./>>45
```
样例输出：
```
3
4
```

3.8 题号：10473

题目描述：输入一个字符串，假定字符串的长度小于等于 256，分别统计出每一种英文字符的个数，不区分大小写。

输入：第一行一个整数 T，表示有 T 组数据。以下 T 行，每行是一个字符串。

输出：对于每组测试数据，先输出"Case #:"（#为序号，从 1 开始），换行，按字母表顺序输出统计结果。数据间用一个空格分隔，行最后一个数据后面无空格。

样例输入：

    2
    ab,cd345!.,/
    abcdefghijklmnopqrstuvwxyz

样例输出：

    Case 1:
    1 1 1 1 0 0 0 0 0 0 0 0 0 0 0 0 0 0 0 0 0 0 0 0 0 0
    Case 2:
    1 1 1 1 1 1 1 1 1 1 1 1 1 1 1 1 1 1 1 1 1 1 1 1 1 1

3.9　题号：10476

题目描述：将一个字符数组 a 中下标为单号的元素赋给另一个字符数组 b，并将其转换成大写字母，然后输出字符数组 b。

输入：第一行是一个整数 T，表示有 T 组数据。以下 T 行，每行是一个字符串，代表字符数组 a，假定长度不超过 256。

输出：对于每组测试数据，输出结果，换行。

样例输入：

    2
    ab,cd345!.,/
    abcdef ghijklmnopqrstuvwxyz

样例输出：

    B D4!,
    BDFGIKMOQSUWY

3.10　题号：10478

题目描述：将字符串 b 连接在字符串 a 的后面，不要使用 strcat 函数。

输入：第一行是一个整数 T，表示有 T 组数据。每组数据占两行，第一行为字符串 a，第二行为字符串 b。两组数据间空一行，假定字符串长度不超过 100。

输出：对于每一组数据，先输出"Case #:"（#为序号，从 1 开始），换行；然后输出连接之后的字符串 a。

样例输入：

    2
    How do
     you do!

    ab,cd345!.,/
    abcdefghijklmnopqrstuvwxyz

样例输出：

Case 1:

How do you do!

Case 2:

ab,cd345!.,/abcdefghijklmnopqrstuvwxyz

3.11 题号：10484

题目描述：编写程序，将输入的一行字符加密和解密。加密时，每个字符依次反复加上"4962873"中的数字，如果范围超过 ASCII 码的 032（空格）~122（"z"），则进行模运算。解密与加密的顺序相反。编制加密与解密函数，打印各个过程的结果。

输入：第一行是一个整数 T，表示有 T 组数据。每组数据一行，为一行字符串，长度不超过 1000 个字符。

输出：对于每一组数据，先输出"Case #:"（#为序号，从 1 开始），换行；然后输出加密的结果，空一行，再输出解密后的结果，换行。

样例输入：

2

By My Side

A Little Too Not Over You

样例输出：

Case 1:

F'&O&'Vmmk

By My Side

Case 2:

E)Rk! oi)Zqw'Qs"&Q#lu$buw

A Little Too Not Over You

3.12 题号：10487

题目描述：编写一个函数，求一个字符串的长度，不能使用 strlen 函数。

输入：第一行是一个整数 T，表示有 T 组数据。每组数据一行，为一行字符串（长度小于等于 255 个字符）。

输出：对于每一组数据，先输出"Case #:"（#为序号，从 1 开始），换行；然后输出字符串长度，换行。

样例输入：

2

By My Side

A Little Too Not Over You

样例输出：

```
Case 1:
10
Case 2:
25
```

3.13 题号：10367

题目描述：

Julius Caesar lived in a time of danger and intrigue. The hardest situation Caesar ever faced was keeping himself alive. In order for him to survive, he decided to create one of the first ciphers. This cipher was so incredibly sound, that no one could figure it out without knowing how it worked.

You are a sub captain of Caesar's army. It is your job to decipher the messages sent by Caesar and provide to your general. The code is simple. For each letter in a plaintext message, you shift it five places to the right to create the secure message (i.e., if the letter is 'A', the cipher text would be 'F'). Since you are creating plain text out of Caesar's messages, you will do the opposite:

Cipher text
A B C D E F G H I J K L M N O P Q R S T U V W X Y Z
Plain text
V W X Y Z A B C D E F G H I J K L M N O P Q R S T U

Only letters are shifted in this cipher. Any non-alphabetical character should remain the same, and all alphabetical characters will be upper case.

输入：

Input to this problem will consist of a (non-empty) series of up to 100 data sets. Each data set will be formatted according to the following description, and there will be no blank lines separating data sets. All characters will be uppercase.

A single data set has 3 components:

Start line - A single line, "START"

Cipher message - A single line containing from one to two hundred characters, inclusive, comprising a single message from Caesar

End line - A single line, "END"

Following the final data set will be a single line, "ENDOFINPUT".

输出：

For each data set, there will be exactly one line of output. This is the original message by Caesar.

样例输入：

START
NS BFW, JAJSYX TK NRUTWYFSHJ FWJ YMJ WJXZQY TK YWNANFQ HFZXJX

END
START
N BTZQI WFYMJW GJ KNWXY NS F QNYYQJ NGJWNFS ANQQFLJ YMFS XJHTSI NS WTRJ
END
START
IFSLJW PSTBX KZQQ BJQQ YMFY HFJXFW NX RTWJ IFSLJWTZX YMFS MJ
END
ENDOFINPUT

样例输出：

IN WAR, EVENTS OF IMPORTANCE ARE THE RESULT OF TRIVIAL CAUSES
I WOULD RATHER BE FIRST IN A LITTLE IBERIAN VILLAGE THAN SECOND IN ROME
DANGER KNOWS FULL WELL THAT CAESAR IS MORE DANGEROUS THAN HE

3.14　题号：11607

题目描述：给定一个32位整数n，将其各位数字剥离。

输入：输入有若干行，每行一个非负整数n。

输出：对每个输入，从高位到低位输出其各位数字，每个数字之间空一格。

样例输入：

```
546
1231
```

样例输出：

```
5 4 6
1 2 3 1
```

第4章　函数

4.1　题号：10452

题目描述：编写一个函数，求三个数中的最大值。

输入：输入有若干行，每行有三个32位整数。

输出：对每一行输入，输出一行，为该行输入的三个数中最大数的数值。

样例输入：

```
2 5 9
8 24 1
```

样例输出：

9
24

4.2 题号：10454

题目描述：编写一个求阶乘的函数。

输入：输入的第一行是一个正整数 m。接下来有 m 行，每行一个整数 n，$1 \leq n \leq 10$。

输出：对每一个 n 输出一行，为 n!。

样例输入：

3
1
5
10

样例输出：

1
120
3628800

4.3 题号：10459

题目描述：本题中的水仙花数是指一个三位正整数，其各位数字的立方和等于该数。例如：407=4*4*4+0*0*0+7*7*7，所以 407 是一个水仙花数。判断给定的数 n 是否为水仙花数。编写一个函数实现判断水仙花数的功能。

输入：输入有若干行，每行一个整数 n，$1 \leq n \leq 999$。

输出：对每一个输入，输出一行，如果是水仙花数，则输出"YES"，否则输出"NO"。

样例输入：

153
100

样例输出：

YES
NO

4.4 题号：10461

题目描述：完数是指不包括其本身的所有因子之和恰好等于其本身的数。例如 6 就是一个完数，因为 6 的所有因子（除了 6 本身之外）是 1、2、3，其和恰好是 6。给定整数 n 判断是否为完数。编写一个函数判断完数。

输入：输入有若干行，每行一个正的 32 位整数 n。

输出：对每个输入输出一行。首先输出"Case 序号:"，然后输出 n 和逗号，再输出"Yes"或"No"。

样例输入：

```
6
9
15
```
样例输出：
```
Case 1: 6,Yes
Case 2: 9,No
Case 3: 15,No
```

4.5 题号：10464

题目描述：素数是只能被 1 和本身整除的整数。例如 2、3、5、7 是素数，4、6、8、9 不是素数。编写一个函数判断给定的整数 n 是否为素数。

输入：第一行是一个正整数 T，其后有 T 行，每行一个正整数 n，n<10000。

输出：对每一个输入 n，输出一行。n 是素数输出 1，否则输出 0。

样例输入：
```
3
2
7
9
```
样例输出：
```
1
1
0
```

4.6 题号：11601

题目描述：n 个数求均值。

输入：输入有若干行，每行第一个数是一个正整数 n（n<1000），其后还有 n 个 32 位整数。编写一个求平均值的函数。

输出：对每一行输入，输出这 n 个数的平均值，保留两位小数。

样例输入：
```
2 3 4
```
样例输出：
```
3.50
```

4.7 题号：11602

题目描述：n 个数求其最小值。

输入：输入有若干行，每行第一个数是一个正整数 n（n<1000），其后还有 n 个 32 位整数。编写一个函数实现。

输出：对每一行输入，输出这 n 个数中最小数的数值。

样例输入：
    2 3 4
样例输出：
    3

4.8　题号：10453

题目描述：有如下分段函数。给定自变量，计算因变量的值。编写一个函数。

$$y=\begin{cases} x & (x<1) \\ 2x-1 & (1\leq x<10) \\ 3x-11 & (x\geq 10) \end{cases}$$

输入：输入第一行是一个正整数 n，其后有 n 行。每行一个数为 x。

输出：根据 x 计算 y 值。如果 y 值是整数，直接输出；否则保留一位小数，四舍五入。

样例输入：
    2
    5
    30
样例输出：
    9
    79

4.9　题号：10454

题目描述：给定如下表达式及其参数，计算表达式的值。编写一个函数。

$$\frac{(x+1)(y-3)}{x+y+z}$$

输入：第一行为一个正整数 n，其后有 n 行，每行三个数，分别是 x、y、z。

输出：对每一组 x、y、z，输出一行为表达式的值。保留四位小数。

样例输入：
    2
    1 2 3
    7 8 20
样例输出：
    -0.3333
    1.1429

4.10　题号：11603

题目描述：编写一个函数判断给定年份是否为闰年。

输入：输入有若干行，每行一个正整数 y，y≤9999。

输出：对每一个输入，输出一行。若 y 是闰年，则输出"Yes"，否则输出"No"。
样例输入：

    1996
    1997

样例输出：

    Yes
    No

4.11　题号：10458

题目描述：计算机中的每一个数都是用二进制表示的。如数值 5，可将其看做是一个 32 位整型数，在计算机中就表示为：0000 0000 0000 0000 0000 0000 0000 0101。一共是 32 个二进制位。

现在将其循环右移一位，将变成：1000 0000 0000 0000 0000 0000 0000 0010。这个数是多少呢？如果是 int 类型，这个数是 -2147483646。

如果将 5 循环右移两位，将变成：0100 0000 0000 0000 0000 0000 0000 0001。这个数就变成 1073741825。

非常奇妙，对 int 做循环移位将使得结果在正负之间来回变换。循环左移的原理是一样的。而且，你应该能够发现把 5 循环左移 31 位，就等于将其循环右移一位。当然这也与 5 用 32 位表示有关，如果用 64 位表示就不是这样了。

输入：第一行为一个正整数 T，其后有 T 行，每行两个 32 位整数 n、m。当 n≥0 时，表示将 m 循环右移 n 位；否则，将 m 循环左移 -n 位。

输出：对每个输入，输出一行，为移位后的结果。

样例输入：

    2
    1 2
    -2 2

样例输出：

    1
    8

4.12　题号：11604

题目描述：很多公式可以用来计算 π，甚至还可以用概率的方法。我们这里提供一个最简单的公式：

$$\pi/4 = 1 - 1/3 + 1/5 - 1/7 + 1/9 - \cdots$$

请利用该公式将 π 计算到指定位数。编写一个函数计算 π 的值。

输入：输入有若干行，每行一个非负整数 n，n 小于等于 12。n 等于 -1 时，测试结束。

输出：对每个输入（-1 除外），输出对应的小数位数的 π，四舍五入。

样例输入：

2
3
-1

样例输出：

3.14

3.142

4.13 题号：11605

题目描述：计算完 π，来计算一下自然对数的底 e 吧。这里有一个简单的公式：

$$e = 1/0! + 1/1! + 1/2! + 1/3! + \cdots$$

编写一个函数计算 e 的值。

输入：输入有若干行，每行一个非负整数 n，n 小于等于 12。n 等于 -1 时，测试结束。

输出：对每个输入（-1 除外），输出对应的小数位数的 e，四舍五入。

样例输入：

2

3

-1

样例输出：

2.72

2.718

第 5 章　指针

5.1 题号：10479

题目描述：定义交换函数 swap，用于交换两个数的值。在主程序中调用 swap 函数。（要求用指针变量传递参数值）。

输入：第一行是一个整数 T，表示有 T 组数据。每组数据一行，为两个整数 a、b。数据间用空格分隔。

输出：对于每一组数据，输出交换后的数据，数据间用一个空格分隔，换行。

样例输入：

2

6 4

4 6

样例输出：

6 4

85 -5

5.2 题号：10480

题目描述：用一个字符指针数组存放所有家庭成员的名单，并把它们打印出来。

输入：第一行是一个整数 T，表示有 T 组数据。每组数据第一行是一个整数 n，表示家庭有 n（n<20）个成员，接下来 n 行为家庭成员的名单（每个成员名字不超过 20 个字符）。

输出：对于每一组数据，先输出"Case #:"（#为序号，从 1 开始），换行；然后输出家庭成员的名单，一个名单一行。

样例输入：

```
2
3
Qu shj
Tan j
Qu yl
2
Liu q
Wang wu
```

样例输出：

```
Case 1:
Qu shj
Tan j
Qu yl
Case 2:
Liu q
Wang wu
```

5.3 题号：10481

题目描述：编写一个程序，向用户询问五种日用品的平均价格，并把它们存放在一个浮点类型的数组中。使用指针从前到后和从后到前的顺序分别打印该数组，然后再用指针把其中的最高价和最低价打印出来。

输入：第一行是一个整数 T，表示有 T 组数据。每组数据一行，五个浮点数，数之间用空格分隔。

输出：对于每一组数据，先输出"Case #:"（#为序号，从 1 开始），换行；然后打印从前到后的顺序，换行；再打印从后向前的顺序，换行；数据之间均用一个空格分隔，每行最后一个数据后面无空格。最后打印 Max:最高价，Min:最低价，换行。

样例输入：

```
2
4 5 8 1 7
12.56 7.38 541.2 79.8 99
```

样例输出：

Case 1:
4 5 8 1 7
7 1 8 5 4
Max:8,Min:1
Case 2:
12.56 7.38 541.2 79.8 99
99 79.8 541.2 7.38 12.56
Max:541.2,Min:7.38

5.4 题号：10482

题目描述：编写一个冒泡排序算法，使用指针将 n 个整型数据按从小到大的顺序进行排序。

输入：第一行是一个整数 T，表示有 T 组数据。每组数据第一个数为 n，表示这组数据有 n 个数，后面紧接着是 n 个数，数之间用空格分隔。

输出：对于每一组数据，先输出 "Case #:"（#为序号，从 1 开始），换行；然后输出排序后的结果，换行；数据之间均用一个空格分隔，每行最后一个数据后面无空格。

样例输入：

2
5 4 5 8 1 7
10 41 67 34 0 69 24 78 58 62 64

样例输出：

Case 1:
1 4 5 7 8
Case 2:
0 24 34 41 58 62 64 67 69 78

5.5 题号：10483

题目描述：编写一个程序将 n 首歌的名字存入一个指针数组中。把这些歌名按原来的顺序打印出来，按字母表的顺序打印出来。

输入：第一行是一个整数 T，表示有 T 组数据。每组数据第一行为整数 n（n<30），表示后面有 n 首歌名，每首歌名一行，歌名不超过 40 个字符。

输出：对于每一组数据，先输出 "Case #:"（#为序号，从 1 开始），换行；然后输出排序前的顺序，空一行，输出排序后的结果，换行；每首歌名一行。

样例输入：

2
3
By My Side

A Little Too Not Over You
Try Again
2
I Need You
Change The World

样例输出：

Case 1:
By My Side
A Little Too Not Over You
Try Again

A Little Too Not Over You
By My Side
Try Again
Case 2:
I Need You
Change The World

Change The World
I Need You

5.6 题号：10484

题目描述：编写程序，将输入的一行字符加密和解密。加密时，每个字符依次反复加上"4962873"中的数字，如果范围超过 ASCII 码的 032（空格）~122（"z"），则进行模运算。解密与加密的顺序相反。编制加密与解密函数，打印各个过程的结果。

输入：第一行是一个整数 T，表示有 T 组数据。每组数据一行，为一字符串，长度不超过 1000 个字符。

输出：对于每一组数据，先输出"Case #:"（#为序号，从 1 开始），换行；然后输出加密后的结果，空一行，再输出解密后的结果，换行。

样例输入：

2
By My Side
A Little Too Not Over You

样例输出：

Case 1:
F'&O&'Vmmk

By My Side

Case 2:
E)Rk! oi)Zqw'Qs"&Q#lu$buw

A Little Too Not Over You

5.7 题号：10485

题目描述：用一个二维数组描述 M 个学生 N 门功课的成绩（假定 M=3，N=4），用行来描述一个学生的 N 门功课的成绩，用列来描述某一门功课的成绩。设计一个函数 minimum 确定所有学生考试中的最低成绩，设计一个函数 maximum 确定所有学生考试中的最高成绩，设计一个函数 average 确定每个学生的平均成绩，设计一个函数 printArray 以表格形式输出所有学生的成绩。

输入：第一行是一个整数 T，表示有 T 组数据。每组数据 M 行为 M 个学生的 N 门功课的成绩，成绩之间用空格分隔。

输出：对于每一组数据，先输出"Case #:"（#为序号，从 1 开始），换行；然后按样例输出样式输出（"No   kc1   kc2   kc3   kc4   average_score"之间空三个空格，之后的所有数据之间空四个空格），每行最后无空格。

样例输入：

1
78 87 67 82
90 67 88 78
80 75 91 81

样例输出：

Case 1:
Min score:67
Max score:91
No   kc1   kc2   kc3   kc4   average_score
1   78   87   67   82   78.5
2   90   67   88   78   80.75
3   80   75   91   81   81.75

5.8 题号：10486

题目描述：使用函数指针数组将题 5.9 的程序改写成使用菜单驱动界面。程序提供五个选项，如下所示（应在屏幕上显示出来）：

Enter a choice :
O   Print the array of grades
1   Find the minimum grade
2   Find the maximum grade
3   Print the average on all tests for each student

4   End program.

输入：第一行是一个整数 T，表示有 T 组数据。每组数据有 M+1 行，前 M 行为 M 个学生的 N 门功课的成绩，成绩之间用空格分隔，M+1 行为一个整数 n，表示选择菜单的次数，后面为 n 个选项（取值 0~4）。

输出：对于每一组数据，先输出"Case #:"（#为序号，从 1 开始），换行；然后按样例输出样式输出（"No   kc1   kc2   kc3   kc4"之间空三个空格，之后的所有数据之间空四个空格），每行最后无空格。

样例输入：
1
78 87 67 82
90 67 88 78
80 75 91 81
5 0 1 2 3 4

样例输出：
```
Case 1:
Enter a choice:
0  Print the array of grades
1  Find the mininum grade
2  Find the maxinum grade
3  Print the average on all tests for each student
4  End program
No   kc1   kc2   kc3   kc4
1    78    87    67    82
2    90    67    88    78
3    80    75    91    81
Min score:67
Max score:91
Average_score:
1    78.5
2    80.75
3    81.75
```

5.9

题目描述：编写一个程序，用随机数产生器建立语句。程序使用四个 char 类型的指针数组 article、noun、verb、preposition。选择每个单词时，在能放下整个句子的数组中连接上述单词。单词之间用空格分开。输出最后的语句时，应以大写字母开头，以圆点结尾。程序产生 20 个句子。

数组填充如下：article 数组包含冠词"the"、"a"、"one"、"some"和"any"，noun 数组包含名词"boy"、"girl"、"dog"、"town"和"car"，verb 数组包含动词"drove"、"jumped"、"ran"、"walked"和"skipped"，preposition 数组包含介词"to"、"from"、"over"、"under"和"on"。

编写上述程序之后，将程序修改成可以产生由几个句子组成的短故事（这样就可以编写一篇自动文章）。

输入：第一行是一个整数 T，表示有 T 组数据。每组数据只有一个整数 n（n≤0），表示产生 n 个句子。

输出：对于每一组数据，先输出"Case #:"（#为序号，从 1 开始），换行；然后输出文章，换行。

样例输入：

2
2
3

样例输出：

Case 1:
A dog skipped to.Any car walked under.
Case 2:
One car drove to.A dog jumped from.The dog ran from.

5.10 题号：10487

题目描述：编写一个函数，求一个字符串的长度，不能使用 strlen 函数。

输入：第一行是一个整数 T，表示有 T 组数据。每组数据一行，为一字符串（长度小于等于 256 个字符）。

输出：对于每一组数据，先输出"Case #:"（#为序号，从 1 开始），换行；然后输出字符串长度，换行。

样例输入：

2
By My Side
A Little Too Not Over You

样例输出：

Case 1:
10
Case 2:
25

第 6 章　结构体与共用体

6.1　题号：10488

题目描述：定义一个结构体变量（包括年、月、日）。计算该日在本年中是第几天。注意闰年问题。

输入：输入数据有若干行。每行上有三个正整数，分别代表年、月、日。

输出：对于每一组数据，输出该日在本年中是多少天。

样例输入：
```
2009 2 20
2009 3 12
```
样例输出：
```
51
71
```

6.2　题号：10489

题目描述：用下列结构描述复数信息。

```
struct complex
{
    int real;
    int im;
};
```

试编写出两个通用函数，分别用来求两复数的和与积。其函数原型分别为：

struct complex cadd(struct complex creal,struct complex cim);
struct complex cmult(struct complex creal,struct complex cim);

即参数和返回值使用结构变量。

输入：输入数据有若干行。每行上有四个整数，前两个数表示一个复数的实部和虚部，后两个数表示另一个复数的实部和虚部。

输出：对于每一组数据，输出两复数的和与积，格式参照样例输出。

样例输入：
```
1 2 3 4
2 1 4 -1
```
样例输出：
```
4+(6i)
-5+(10i)
6+(0i)
9+(2i)
```

6.3 题号：10490

题目描述：输入 n 本书的名称、单价、作者和出版社，按书名升序进行排版和输出（长度均不超过20个字符）。

输入：输入数据有若干组。每组数据第一行是一个整数 n，表示有 n 本书，接下来每一行是一本书的信息（名称 单价 作者 出版社）。

输出：对于每一组数据，输出排序后的结果，输出格式采用左对齐（cout.setf(ios::left)）和 setw(20)，每组数据后空一行。

样例输入：

```
2
Datastructure 30 Wangweiguang Hunan_University
C++ 26 Liuhong Wuhan_University
3
Hongloumeng 40 Caoxueqing RenMing
Computer_network 50 Liug Hunan_normal
Frontpage 23 Qshj Tsinghua
```

样例输出：

```
C++                 26    Liuhong             Wuhan_University
Datastructure       30    Wangweiguang        Hunan_University

Computer_network    50    Liug                Hunan_normal
Frontpage           23    Qshj                Tsinghua
Hongloumeng         40    Caoxueqing          RenMing
```

6.4 题号：10491

题目描述：幼儿园老师组织小朋友进行游戏，13个人围成一圈，从第1个人开始顺序报号1、2、3。凡报到"3"者退出圈子。请你帮小朋友找出最后留在圈子的人原来的序号。我们这里假设报数的人数为 N（1≤N≤200）。

输入：第一行为一个整数 T（1≤T≤100），表示共有多少组测试数据。接着下面的 T 行数据为 T 组报数的人数 N（1≤N≤200）。

输出：对每组测试数据，第一行输出小朋友离开的次序，第二行输出最后留在圈子的人原来的序号。

样例输入：

```
2
3
5
```

样例输出：

```
Sequence that persons leave the circle:
3 1
The last one is:2
```

```
Sequence that persons leave the circle:
3 1 5 2
The last one is:4
```

6.5 题号：10492

题目描述：已知枚举类型定义如下：

enum color{red, yellow, blue, green, white, black};

从键盘输入一整数，显示与该整数对应的枚举常量的英文名称。

输入：输入数据有若干行，每行一个整数。

输出：对每行数据，输出与该整数对应的枚举常量的英文名称，换行。

样例输入：

```
0
1
```

样例输出：

```
red
yellow
```

第7章 类与对象及封装性

7.1 写出下列程序的运行结果。

```
#include<iostream.h>
class Cat
{
public:
    int GetAge( );
    void SetAge(int age);
    void Meow( );
protected:
    int itsAge;
};
int Cat::GetAge( )
{
    return itsAge;
}
void Cat::SetAge(int age)
{
```

```
        itsAge=age;
}
void Cat::Meow( )
{
        cout<<"Meow.\n";
}
int main( )
{
        Cat frisky;
        frisky.SetAge(5);
        frisky.Meow( );
        cout <<"frisky is a cat who is"<<frisky.GetAge()<<"years old.\n";
        frisky.Meow( );
        return 0;
}
```

7.2 题号：10495

题目描述：用类实现求两复数的加法和乘法。

输入：输入数据有若干行。每行上有四个数，前两个数表示一个复数的实部和虚部，后两个数表示另一个复数的实部和虚部。

输出：对于每一组数据，输出两复数的和与积，格式参照样例输出。

样例输入：

```
1 2 3 4
2 1 4 -1
```

样例输出：

```
4+(6i)
-5+(10i)
6+(0i)
9+(2i)
```

7.3 题号：10496

题目描述：定义一个满足如下要求的 Date 类。

（1）用下面的格式输出日期：日/月/年

（2）可运行在日期上加一天操作；

（3）设置日期。

输入：输入数据有若干组。每组数据一行，有三个整数，表示日期，格式为：日 月 年。

输出：对于每一组数据，输出日期，日期加一天后，再输出日期，均需要用类的成员函数实现。

样例输入：
    20 1 2010
    31 12 2002
    29 2 2008

样例输出：
    20/1/2010
    21/1/2010
    31/12/2002
    1/1/2003
    29/2/2008
    1/3/2008

第 8 章　类的深入

8.1　通过上机实践，分析下列程序所输出的结果。

```
#include<iostream>
using namespace std;

class myclass
{
    int *p;
public:
    myclass(int i);//构造函数
    myclass(const myclass &ob);//拷贝构造函数
    ~myclass();//构造函数
    void Show(){ cout<<*p;}
};
myclass::myclass(int i)
{
    cout<<"Normal constructor\n";
    p = new int;
    if(!p)
    {
        cout<<"Allocation failure\n ";
```

```cpp
        exit(1);
    }
    *p = i;
}
myclass::myclass(const myclass &obj)
{
    cout<<"Copy constructor\n";
    p = new int;
    if(!p)
    {
        cout<<"Allocation failure\n ";
        exit(1);
    }
    *p = *obj.p; //copy value
}
myclass::~myclass()
{
    cout<<"Destructed\n";
    delete p;
}
void display(myclass ob)
{
    ob.Show();
}

int main( )
{
    myclass a(10);
    display(a);
    return 0;
}
```

8.2 题号：10497

题目描述：用类和友元函数来实现复数的加法和乘法。

输入：输入数据有若干行。每行上有四个数，前两个数表示一个复数的实部和虚部，后两个数表示另一个复数的实部和虚部。

输出：对于每一组数据，输出两复数的和与积，格式参照样例输出。

样例输入：

```
1 2 3 4
2 1 4 -1
```
样例输出:
```
4+(6i)
-5+(10i)
6+(0i)
9+(2i)
```

8.3 题号：10498

题目描述：设计一个日期类 Date，包括日期的年份、月份和日号，编写一个友元函数，求两个日期 d1、d2 之间相差的天数 d2-d1。

输入：输入数据有若干组。每组数据一行，有六个整数，表示两个日期 d1、d2，格式为：年 月 日。

输出：对于每一组数据，输出两个日期 d1、d2 之间相差的天数，格式参照样例输出。

样例输入：
```
2000 1 1 2002 10 1
2010 1 10 2010 3 20
```
样例输出：
```
2002/10/1-2000/1/1=1004
2010/3/20-2010/1/10=69
```

8.4 题号：11608

题目描述：对象也能作为函数的参数，但是这个过程并不是这么简单。请采用对象传入函数的方式来计算三角形的面积，此外，对这个三角形类设置其构造函数和析构函数，且构造函数中应输出 Constructing，析构函数中应输出 Destructed，对于求面积的函数，请采用普通函数。

输入：每次输入三个浮点数 a,b,c，分别表示三角形的三条边的长度，输入都能构成一个三角形。

输出：对于每一行的输入，在下一行输出中间过程所输出的字符串面积，面积保留两位小数。

样例输入：
```
3 4 5
```
样例输出：
```
Constructing
Destructed
6.00
Destructed
```

8.5　题号：11609

题目描述：函数的返回值也能是一个对象。请采用类的方式来编写一个关于坐标点绕原点旋转一定角度，得到新的坐标点，要求编写一个类名为 point 的类表示坐标点，实现一个普通函数来计算新坐标点，且该函数的返回值类型是对象 point。

输入：每次输入三个浮点数 x,y,a，分别表示这个点的横坐标与纵坐标，a 表示这个坐标点绕原点逆时针方向旋转的角度。

输出：对于每一行的输入，在下一行输出新点的横坐标与纵坐标，保留两位小数。

样例输入：

　　1 0 90

样例输出：

　　0.00 1.00

8.6　题号：11610

题目描述：请用友元函数和拷贝构造函数一起来实现这样一个问题：给出三角形的三个坐标，求这个三角形的面积。在拷贝构造函数中输出 copy constructor，友元函数中实现求面积过程。

输入：每次输入六个浮点数 x1,y1,x2,y2,x3,y3，分别表示三个坐标 A(x1,y1)，B(x2,y2)，C(x3,y3)。

输出：对于每一行的输入，在下一行输出拷贝构造函数中的字符串和面积，并保留两位小数。

样例输入：

　　0 0 1 0 0 1

样例输出：

```
copy constructor
copy constructor
copy constructor
0.50
```

第 9 章　运算符重载

9.1　题号：10499

题目描述：定义复数类的加法与减法，使之能够执行下列运算。

```
Comples a(2,5),b(7,8),c(0,0);
c=a+b;
c=4.1+a;
```

c=b+5.6;

输入：输入数据有若干组。每组两行，第一行上有四个数，前两个数表示复数 a 的实部和虚部，后两个数表示复数 b 的实部和虚部。第二行为两个实数 m、n。第一个实数 m 为求 m+a，第二个实数 n 为求 b+n。

输出：对于每一组数据，分别输出求和后的结果，格式参照样例输出。

    a+b

    m+a

    b+n

样例输入：

    2 5 7 8

    4.1 5.6

    1 2 3 4

    4.5 6.3

样例输出：

    9+(13i)

    6.1+(5i)

    12.6+(8i)

    4+(6i)

    5.5+(2i)

    9.3+(4i)

9.2 题号：10500

题目描述：编写一个时间类，实现时间的加、减、读和输出。

输入：输入数据有若干组。每组数据一行，有六个整数，表示两个日期 d1,d2，格式为：年 月日。

输出：对于每一组数据，输出两个日期 d1,d2 之间相差的天数，格式参照样例输出。

样例输入：

    15 5 10 5 45 40

    20 12 50 10 55 40

样例输出：

    15:5:10+5:45:40=20:50:50

    15:5:10-5:45:40=9:19:30

    20:12:50+10:55:40=7:8:30

    20:12:50-10:55:40=9:17:10

9.3 题号：10502

题目描述：设计一个分数类 rationalNumber，该类中包括分子和分母两个成员数据，并具有下述功能。

（1）建立构造函数，它能防止分母为零（分母为 0 时，输出"denominator equal zero"），当分数不是最简形式时进行约分，并避免分母为负数。

（2）重载加法运算符。

（3）重载关系运算符：>、<、==。

输入：输入数据第一行为一个整数 T，表示有 T 组数据。每组数据一行，包含四个整数，分别表示两个分数 a,b 的分子和分母：分子 分母 分子 分母。

输出：对于每一组数据，分别输出两个分数相加，关系运算>、<、==的结果，格式参照样例输出。

样例输入：

2

2 -4 1 4

2 0 4 6

样例输出：

-1/2+1/4=-1/4

-1/2>1/4=0

-1/2<1/4=1

-1/2==1/4=0

denominator equal zero

第 10 章 继承性

10.1 调试下列程序代码，并分析它的编译出错信息。

```
#include<iostream>
using namespace std;
class base
{
    protected:
    int i,j;
    public:
    void set(int a,int b)
    {
        i = a ;
        j = b;
    }
    void show()
```

```
        {
            cout<<i<<" "<<j<<"\n";
        }
};
class derived1:private base
{
    int k;
public:
    void setk()
    {
        k = i*j;
    };
    void showk()
    {
        cout<<k<<"\n";
    };
};
class derived2: public derived1
{
    int m;
public:
    void setm()  { m = i-j; };
    void showm() { cout<<m<<"\n"; };
};

int main()
{
    derived1 ob1;
    derived2 ob2;
    ob1.set(1,2);
    ob1.show();
    ob2.set(3,4);
    ob2.show();
    return 0;
}
```

10.2 题号：10503

题目描述：建立一个基类 Building，用来存储一座楼房的层数、房间数以及它的总平方英尺数。建立派生类 Housing，继承 Building，并存储下面的内容：卧室和浴室的数量，另建立派生类 Office，继承 Building，并存储灭火器和电话的数目。然后，编制应用程序，建立住宅楼对象和办公楼对象，并输出它们的有关数据。

输入：输入数据第一行为一个整数 T，表示有 T 组数据。每组数据两行，每行包括五个数。

第一行：层数 房间数 总平方英尺数 卧室数 浴室数
第二行：层数 房间数 总平方英尺数 灭火器数 电话数
输出：对于每一组数据，分别输出有关数据。格式参照样例输出。
样例输入：

1
4 8 240.45 2 2
8 12 500.5 12 2

样例输出：

HOUSING:
Floors:4
Rooms:8
Total area:240.45
Bedrooms:2
Bathrooms:2
OFFICING:
Floors:8
Rooms:12
Total area:500.5
Extinguishers:12
Phones:2

10.3 题号：11611

题目描述：富士康的张全因为英语太好，被分配到管理 iPhone 手机。iPhone 有很多版，有 S 和 C 等众多款，而且每一款都有各种颜色，各个版本也有不同的价格。公司现在生产 iPhone 5 和 iPhone 5S。让张全制作一张报价表，他觉得 so easy，就交给了你。于是由你完成这个任务，要求你的代码中 iPhone 5S 类需要继承 iPhone 5 类，否则张全会扣你的工资。

输入：第一行包含一个正整数 T（0 < T ≤ 10）表示有 T 组测试数据。

对于每组测试数据占两行，第一行为生产的 iPhone 5 的颜色和价格，第二行为 iPhone 5S 的信息。对于每一行颜色为首字母大写，其他均为小写字母并且长度不超过 10 个字符，价格为不超过 10 位的正整数，价格和颜色中间由空格隔开。

输出：对于每一组测试数据，第一行为 Case #t:，t 表示当前为第 t 组，从 1 开始。第二行为 iPhone 5 的信息，格式为 iPhone 5 { Color : color / Prize : prize }，第三行为生产的

iPhone 5s 的信息，格式为 iPhone 5S { Color :color / Prize : prize }。其中 color 为输入对应的颜色，prize 为输入对应的价格。

样例输入：
```
3
Diaosibai 123456789
Tuhaojin 987654321
Hahahahaha 5201314
Hehe 66666
Wqnmlgb 233333
Malatang 6
```

样例输出：
```
Case #1:
Iphone5 { Color : Diaosibai \ Prize : 123456789 }
Iphone5S { Color : Tuhaojin \ Prize : 987654321 }
Case #2:
Iphone5 { Color : Hahahahaha \ Prize : 5201314 }
Iphone5S { Color : Hehe \ Prize : 66666 }
Case #3:
Iphone5 { Color : Wqnmlgb \ Prize : 233333 }
Iphone5S { Color : Malatang \ Prize : 6 }
```

10.4 题号：11612

题目描述：老罗的锤子手机简直是国产神机，小明买了一台锤子，但是他想知道他买的这台组装机值不值这个价，你身为张全手下的质检员肯定相当了解。如果一台锤子的组装零件的价格不小于这台手机价格的百分之八十七，那么就觉得买值了。你现在可以编写一个程序，其中有四个类，分别为 Sony、iPhone、Mi、Chuizi，要求 Chuizi 继承前三个类。每个类都包含一个价格成员，然后比较 Chuizi 的价格和前三个的价格和，并且告诉小明，买锤子值不值。

输入：第一行包含一个整数 T（0<T≤100），接下来为 T 组测试数据，其中每组数据占一行，包含四个正整数 v1、v2、v3、v4。V1 为 Sony 的价格，V2 为 iPhone 的价格，V3 为 Mi 的价格，V4 为锤子的价格。所有价格都保证小于 2^31。

输出：对于每组测试数据，输出占一行，为 Case #t: ans。其中 t 表示第 t 组测试数据，ans 为 "Yes" 表示买赚了或者 "No" 表示买亏了，t 从 1 开始。

样例输入：
```
3
10 20 30 100
40 40 40 100
87 0 0 100
```

样例输出：
```
Case #1: No
Case #2: Yes
Case #3: No
```

第 11 章　多态性

11.1　题号：11613

题目描述：有一所学校的某年级有两个班：class1 和 class2。最近该年级进行了一次数学考试（满分为 100 分）。现给你该次考试 class1 的全班总分和平均分以及全年级的总分和平均分，试求 class2 的平均分。

要求：建立一个班类，计算函数放在类中定义。调用函数要求使用指针，并要求定义的指针数量尽量少。

输入：只有一组测试数据，该组测试数据包括两行：每行两个数字 a 和 b，第一行表示 class1 的全班总分和平均分，第二行表示全年级的总分和平均分。所有输入保证是整数并且范围在 1~10000 之间。

输出：输出一个数字 class2 的平均分即可。

（所有输入数据保证 class2 的平均分最后计算出来的也是正整数。）

样例输入：
```
1000 50
2000 50
```
样例输出：
```
50
```

11.2　题号：11614

题目描述：给出一个坐标点的坐标，求这个坐标点到原点的距离。给出的坐标可能是二维坐标也可能是三维坐标。

要求：定义点（point）类；计算二维坐标和三维坐标距离的函数要同名（用虚函数实现）。

输入：第一行输入一个正整数 t（1≤t≤10），代表测试数据组数。从第二行开始每一组测试数据。每组测试数据一行：首先是一个小写字母 b 或 t，b 表示该坐标是二维坐标，后面接着该点的坐标 x,y。t 表示该坐标是三维坐标，后面接着该点坐标 x,y,z。所有坐标均为整数，且绝对值不超过 100。

输出：对输入的每个点，输出其到原点的距离，结果保留两位小数。

样例输入：
```
2
```

```
    b 1 1
    t 2 2 2
```
样例输出：
```
    1.41
    3.46
```

11.3　题号：11615

题目描述：阅读下面的程序，现给多组测试数据，你的任务是将程序运行的结果输出来。

注意：此题直接提交原程序可能超时。

```cpp
    #include<iostream>
using namespace std;
class Base
{
    public:
            virtual void opt() { cout << "Apple" << endl; }
};
class cs1 : public Base
{
    public:
            void opt() { cout << "Banana" << endl; }
};
class cs2 : public Base
{
    public:
            //void opt() { cout << "Car" << endl; }    注意此行已被注释
};
int main()
{
    int t;
    Base A;
    cs1 B;
    cs2 C;
    while ( cin >> t && t )
    {
        if(t == 1)
        {
            Base *p;
```

```
                p = &A;
                p -> opt();
            }
            else if (t == 2)
            {
                cs1 *p;
                p = &B;
                p -> opt();
            }
            else
            {
                cs2 *p;
                p = &C;
                p -> opt();
            }
        }
        return 0;
    }
```

输入：测试数据有多组，对于每组测试数据，输入 t（t=0,1,2,3）。

输出：输出该程序运行的结果即可。

样例输入：

   1
   2
   0

样例输出：

   Apple
   Banana

11.4 题号：11616

题目描述：给定一个二维坐标点，求该点到原点的二维距离和直角距离。

提示：点 A(x1,y1)、B(x2,y2)之间直角距离 d=|x1-x2|+|y1-y2|。

程序中必须包括下面的内容：

```
class point
{
    protected:
        int x,y;
    public:
        void set_point(int a,int b) { x = a,y = b; }
```

```
virtual void dis() {cout << "Undefinition." << endl; }
};
```

输入：第一行输入一个整数 t（1≤t≤100），代表测试数据组数。每组数据一行，该行包括两个整数 a、b，表示一个直角坐标系下的坐标（-100≤a,b≤100），中间用空格隔开。

输出：对于每组测试数据，输出该点到原点的二维距离和直角距离，每个一行。二维距离保留两位小数。

样例输入：

  2
  1 1
  3 4

样例输出：

  1.41
  2
  5.00
  7

第 12 章　输入/输出流

12.1　编写程序，读入一个文件，统计文件中的行数。
12.2　把一个文件的内容复制到另一个文件中。
12.3　将结构体类型的数据写入一个二进制文件中。
12.4　建立 RMB 类，重载插入运算符，使得人民币输出格式为：长度一律 10 位，小数两位，币值前有一￥符号。

注意：本章题目涉及文件操作，未录入 OJ 中，请自己在计算机上完成练习。

第 13 章　模板和标准库

13.1　题号：10543

题目描述：FJ is surveying his herd to find the most average cow. He wants to know how much milk this 'median' cow gives: half of the cows give as much or more than the median; half give as much or less.

Given an odd number of cows N (1≤N<10,000) and their milk output (1..1,000,000), find the median amount of milk given such that at least half the cows give the same amount of milk or more and at least half give the same or less.

输入：Line 1: A single integer N。Lines 2..N+1: Each line contains a single integer that is the milk output of one cow.

输出：Line 1: A single integer that is the median milk output.

样例输入：

5
2
4
1
3
5

样例输出：

3

提示：使用 partial_sort()算法。

13.2　题号：10595

题目描述：历史书上说，自统治者 Big Brother 去世以后，大洋国就陷入了无休止的内战中，随时可能有新的武装势力出现，随时可能有战争发生。奇怪的是，每次战争都是在当前国内战斗力最强大的两股势力间进行，我们可以把每股武装势力的战斗力量化成一个值，每次战争都是在当前战斗力值最高的两股势力间进行。如果有多支势力战斗力值相同，则名字字典序大的在前（见下面第二组样例）。一场战争结束后，战斗力稍弱的那方被消灭，另一方也元气大伤，战斗力减弱为两支武装的战斗力之差。如果发生战争的两方战斗力相同，则他们会同归于尽。历史书上详细记录了该段时期的事件，记录分为两种格式：

（1）New name value，其中 name 和 value 是变量，表示一个名字叫做 name，战斗力为 value 的新势力出现。

（2）Fight 表示在当前最强的两股势力间发生了战争。

现在请你根据书上记录，计算出每场战争以后分别导致哪支（或哪两支）势力被消灭。

输入：输入的第一行包含一个整数 T（T≤15），表示共有 T 组数据。接下来每组数据的第一行是一个整数 N（N≤50000），表示有 N 条记录。接下来 N 行，每行表示一条记录，记录的格式如上所述。输入保证每股势力的名字都不相同，势力的名字仅包含小写字母，长度不超过 20 个。战斗力值为不超过 10000 的正整数。保证当战争发生时至少有两支势力存在。

输出：对每组数据，输出一行"Case X:"作为开头，此处 X 为从 1 开始的编号。注意首字母 C 为大写，在"Case"和编号 X 之间有一个空格，在编号 X 后面有一个冒号。然后对每条 Fight 记录输出一行，表示被消灭的势力的名字。如果是两支势力同归于尽，则这两个名字都应该输出，字典序大的在前，两个名字之间用一个空格隔开。

样例输入：
```
2
5
New obrien 100
New winston 199
Fight
New julia 99
Fight
4
New miniluv 100
New minipax 100
New minitrue 100
Fight
```

样例输出：
```
Case 1:
obrien
winston julia
Case 2:
minitrue minipax
```

提示：使用 set 容器。

13.3　题号：10895

题目描述：DD 同学刚刚开始学习数据结构，老师希望他能运用计算机进行建档的集合运算（并、交、差），但是 DD 同学觉得他才刚刚学完线性表，想了很久都不会做，而老师说了只要用所学的知识就能解决，聪明、好学的你能帮 DD 同学解决这个问题吗？

输入：输入文件有若干组；每组测试数据包括两行字符串，分别代表两个集合 A 和 B，其中的字母限定为小写字母；字符串是连续的（即中间不包含其他字母）；字符串中前面出现的字母的 ASCII 码都要比后面的小且不会出现相同的字母。

输出：每一组对应的输出数据包括四行，第一行输出 Case n:，以后的后面三行每行一个字符串，依次对应为 A 与 B 的并、A 与 B 的交、A 与 B 的差（即 A−B），输出的字符串要保证后面出现的字符比前面出现的字符 ASCII 码大。

样例输入：

abc

cdf

样例输出：

Case 1:

abcdf

c

ab

提示：使用 set 容器，配合集合运算算法 set_union()、set_intersection()、set_difference()。

13.4 题号：10898

题目描述：这是一个既经典又简单的题目，题目描述如此简单，给你两个有序的升序整数序列，将其合并后输出。

输入：输入数据有若干组，单个输入文件只有一组数据。每组测试数据包括四行：第一行为整数 M，代表第一个序列的长度；第二行输入第一个序列的 M 个数，每两个数之间用空格隔开；第三行输入整数 N，代表第二个序列的长度；第四行输入第二个序列的 N 个数，每两个数之间用空格隔开。其中 $0 \leq M, N \leq 2,000,000$，每个序列的整数都不超过 100000，不小于 0。

输出：升序输出合并之后的有序序列，重复的数字都要输出，每两个数之间用空格隔开，最后一个数后面不含空格，但要换行。

样例输入：

2

1 3

1

2

样例输出：

1 2 3

提示：使用 merge()算法，或者 inplace_merge()算法。

13.5 题号：10919

题目描述：zyl 公司搞了一次重大的无聊的年考，一共有 n（$1 \leq n \leq 1000000$）个员工参加了考试，每个参与者又考了两门，一门是数据结构，另一门是中国古代史。z 老板想知道考试后员工的排名（数据结构，中国古代史，优先级由高到低）。a 员工很粗心，只排好了 n−1 个人的成绩，现在 z 老板要你告诉他第 n 个员工在 n−1 个人中的名次（名次最靠前）。

输入：输入只有一组数据。第一行一个数 n 表示参与考试的有 n（$1 \leq n \leq 1000000$）个员工（前 n−1 个员工是乱序的）。第二行到第 n 行每行两个数，第一个数是数据结构的分数，第二个数是中国古代史的分数。第 n+1 行为一个数 t（1≤t≤10000），表示第 n 个员工的成绩有 t 钟可能。接下来 t 行，每行两个数，第一个数是数据结构的分数 d，第二个数是中国古代史的分数 h，表示第 n 个员工此次考试的成绩（成绩在 0≤d，h≤100）。

输出：对于第 n 个员工的成绩，输出这个员工的成绩排名（名次最靠前的），一行一个数。

样例输入：
3
70 70
80 80
3
70 70
70 75
80 80

样例输出：
2
2
1

提示：使用 sort() 算法。

13.6　题号：11068

题目描述：此题既容易又能让你放松心情，有安定神闲，舒筋脉络，气血畅通的作用，紧张的，心急的，抢速度的，想抢 FB 的，还没 A 题的，此题是改善心情，放松身体之良药。

给你有序的几个数字（不超过 20 个），每个数字不大于 16，里面会有些重复的，把它们去掉，然后输出。

输入：每组测试数据分两行：第一行为 n（1≤n≤20），表示数字的个数；第二行为 n 个数 A1，A2，…，An。测试数据以 n=0 为结束。

输出：对于每一组测试数据，输出一行，输出去掉重复数字后的序列。每行结尾无空格。

样例输入：
5
1 1 6 9 12
4
0 1 7 10
0

样例输出：

```
1  6  9  12
0  1  7  10
```

提示：使用 unique()算法。

13.7　题号：11208*

题目描述：Buy low, sell high. That is what one should do to make profit in the stock market (we will ignore short selling here). Of course, no one can tell the price of a stock in the future, so it is difficult to know exactly when to buy and sell and how much profit one can make by repeatedly buying and selling a stock.

But if you do have the history of price of a stock for the last n days, it is certainly possible to determine the maximum profit that could have been made. Instead, we are interested in finding the k1 lowest prices and k2 highest prices in the history.

输入：The input consists of a number of cases. The first line of each case starts with positive integers n, k1, and k2 on a line ($n \leqslant 1,000,000$, $k1 + k2 \leqslant n$, $k1, k2 \leqslant 100$). The next line contains integers giving the prices of a stock in the last n days: the i-th integer ($1 \leqslant i \leqslant n$) gives the stock price on day i. The stock prices are non-negative. The input is terminated by n = k1 = k2 = 0, and that case should not be processed.

输出：For each case, produce three lines of output. The first line contains the case number (starting from 1) on one line. The second line specifies the days on which the k1 lowest stock prices occur. The days are sorted in ascending order. The third line specifies the days on which the k2 highest stock prices occur, and the days sorted in descending order. The entries in each list should be separated by a single space. If there are multiple correct lists for the lowest prices, choose the lexicographically smallest list. If there are multiple correct lists for the highest prices, choose the lexicographically largest list.

样例输入：
```
10 3 2
1 2 3 4 5 6 7 8 9 10
10 3 2
10 9 8 7 6 5 4 3 2 1
0 0 0
```

样例输出：
```
Case 1
1 2 3
10 9
Case 2
8 9 10
```

2 1

提示：使用 multiset 容器，以及 sort()算法。

13.8 题号：11294

题目描述：给出一系列数据，请你剔除其中重复的数据。本题判断重复的功能，要求编写一个函数。整个程序不能用全局变量和全局数组。

输入：有多组测试数据。输入的第一行是整数 T（0<T≤100），表示测试数据的组数。每一组测试数据只有一行，第一个是整数 n，表明随后有 n 个整数；接着其后有 n 个整数，该行每个数后均有一个空格。该行没有其他多余的符号。1<n≤10^4，数据的范围为[-5000,5000]。

输出：对应每组输入，输出一行结果，第一个数说明非重复元素的个数，随后是非重复的元素（按从小到大），该行每个数后应有一个空格。该行不能有其他多余的符号。

样例输入：

1
10 9 8 6 7 5 4 5 9 5 7

样例输出：

6 4 5 6 7 8 9

提示：使用 sort()算法以及 unique()算法。

13.9 题号：11298

题目描述：我们知道，每一年的 ACM-ICPC 亚洲地区现场赛都会有很多个代表队参加，于是，如何排序这些队伍成了组织者最头疼的问题。后来，组织者终于想出了解决方案，那就是按照网络赛的做题数排序。如果做题数一样，就按罚时排序；如果罚时一样，就按照学校名字的字典序排序。

组织者规定：

（1）做题数多的排在前面。

（2）罚时少的排在前面。

（3）名字按字母 ASCII 码值升序排列，如果第一个字母一样，就比较第二个字母，以此类推。

输入：Line 1：一个数 N（1≤N≤500），代表进入现场赛的学校数。Line 2~Line N+1：队伍的名字（不含空格，由大小写字母组成，长度不超过 50 个）、做题数（正整数）、罚时（正整数）。输入数据保证不出现三者完全一样的数据。

输出：按输入的格式将队伍排序后输出，每个队伍一行。

样例输入：

2
TeamA 3 200

　　　　TeamB 4 500
　样例输出：
　　　　TeamB 4 500
　　　　TeamA 3 200
　提示：使用 sort()算法。

13.10　题号：11617

　题目描述：在湖南师范大学上学的人中，总有些大神（或许也是无聊的人），这天，王二说他发明了一种新的文字，声称可以和火星人交流。当然，王二不是那么肤浅的人，告诉你只是想炫耀一下，他已为这种文字和英文之间创立了字典，他想让你对照这个字典把一篇英语文章翻译成王二文（即他创立的新文字），虽然王二知道你很不情愿，但他告诉你，如果你翻译完就给你介绍女朋友。

　输入：问题只有一组测试样例，输入包含两部分，字典部分和翻译的文章部分，第一行为 START，表示字典部分输入开始，字典部分每行有两个字符串，第一个字符串为英文，第二个字符串为所对应的王二文，字典部分的最后一行为 END，表示字典部分结束。接下来为文章部分，文章部分以 START 开始 END 结束，START、END 均不用翻译，如果某个英文单词在字典中没有所对应的王二文，就输出原单词。注意文章可能有多行和多个空格，要按原文章的格式输出。

　在本问题中，字典部分和文章部分所涉及的英文字母均为小写，且每个单词不超过 10 个字母。

　输出：最后你所翻译完成的文章。

　样例输入：
　　　　START
　　　　is shi
　　　　END
　　　　START
　　　　wanger is sb.
　　　　END
　样例输出：
　　　　wanger shi sb.

第三部分　参考答案

第1章 C++语言概述

1.1 参照第一部分内容第 1.1~1.4 节。
1.2 参照第一部分内容第 1.1~1.4 节。
1.3 参照第一部分内容第 1.1~1.4 节。
1.4 参照第一部分内容第 1.5 节。
1.5 题号 10000

```cpp
#include <iostream>
using namespace std;
int main()
{
    int a, b;
    cin >> a >> b;
    cout << a + b << endl;
    return 0;
}
```

补充知识点：

<stdio.h>是 C 语言中的标准库输入/输出头文件，使用 C 中的输入函数 scanf()，输出函数 printf()就要包含此头文件。

<iostream>和<iostream.h>是 C++中的标准库输入/输出数据流文件，两者是不一样的，前者没有后缀，实际上，在你的编译器 include 文件夹里可以看到，两者是两个文件，打开文件就会发现，里面的代码是不一样的。后缀为.h 的头文件 C++标准已经明确提出不支持了，早些的实现将标准库功能定义在全局空间里，声明在带.h 后缀的头文件里，C++标准为了和 C 区分开，也为了正确使用命名空间，规定头文件不使用后缀.h。因此，当使用<iostream.h>时，相当于在 c 中调用库函数，使用的是全局命名空间，也就是早期的C++实现；当使用<iostream>时，该头文件没有定义全局命名空间，必须使用 namespace std; 这样才能正确使用 cout。

因此，在 Microsoft Visual C++ 6.0 编程环境中可以使用如下两种形式。

```cpp
#include<iostream>
using namespace std;
void main()
{
    cout<<"Hello world!"<<endl;
```

}

或者：

```
#include<iostream.h>
void main()
{
    cout<<"Hello world!"<<endl;
}
```

而在 Microsoft Visual Studio 2005 中只能使用如下形式：

```
#include<iostream>
using namespace std;
void main()
{
    cout<<"Hello world!"<<endl;
}
```

第 2 章　C++语言编程基础

2.1　题号：10452

求三个整数中最大的数。

```
#include<iostream>
using namespace std;
int main()
{
    long a,b,c,max;
    while(cin>>a>>b>>c)  //从键盘读入 a,b,c 三个数
    {
        max=a;  //先假定 a 最大
        if(a<b)  //a,b 比较，求 a,b 中大数
            max=b;
        if(max<c)  //求最大数
            max=c;
        cout<<max<<endl;
    }
    return 0;
}
```

2.2 题号：10454

求 n 的阶乘。

```cpp
#include<iostream>
using namespace std;
int main()
{
    long m,n;
    long i,jc;
    cin>>m;
    while(m--)
    {
        cin>>n;
        jc=1;
        for(i=1;i<=n;i++)
        {
            jc*=i;
        }
        cout<<jc<<endl;
    }
    return 0;
}
```

2.3 题号：10459

求水仙花数。

```cpp
#include<iostream>
using namespace std;
int main()
{
    int n,g,s,b;//n为要移的位数，m为要移位的数
    while(cin>>n)
    {
        b=n/100;      //取出三位数的百位
        s=n/10%10;    //取出三位数的十位
        g=n%10;       //取出三位数的个位
        if(n==b*b*b+s*s*s+g*g*g)
            cout<<"YES"<<endl;
```

```
        else
            cout<<"NO"<<endl;
    }
    return 0;
}
```

2.4 题号：10461

完数。

```cpp
#include<iostream>
using namespace std;
int main()
{
    int i=0,j,sum,n;  //定义数,因子
    while(cin>>n)
    {
        sum=0;
        i++;
        for(j=1;j<n;j++)
        {
            if(n%j==0)  //当i被j整除
            {
                //原数等于除完因子后的数
                sum+=j;  //把因子加到sum中
            }
        }
        if(sum==n)  //因子相加和等于原数i时
            cout<<"Case "<<i<<": "<<n<<",Yes"<<endl;
        else
            cout<<"Case "<<i<<": "<<n<<",No"<<endl;
    }
    return 0;
}
```

2.5 题号：10464

素数。

```cpp
#include<iostream>
#include<cmath>
```

```cpp
using namespace std;

int main()
{
    int i,T,n;
    cin>>T;
    while(T--)
    {
        cin>>n;
        int flag=1;
        for( i=2;i<=int(sqrt(n)); i++)
        {
            if ( n%i == 0 )
            {
                flag=0;   //n不是素数
                break;
            }
        }
        if(flag)
            cout<<1<<endl;
        else
            cout<<0<<endl;
    }
    return 0;
}
```

2.6 题号：11601

求平均值。

```cpp
#include <iostream>
#include <iomanip>
using namespace std;
typedef long long llt;
int main(){
    int n;
    while ( cin >> n ){
        llt s = 0LL;
        for(int i=0;i<n;++i){
```

```
        llt x;
        cin >> x;
        s += x;
    }
    cout<<fixed<<setprecision(2)
        <<(double)s/(double)n<<endl;
}
return 0;
}
```

2.7 题号：11602

求最小值。

```
#include <iostream>
using namespace std;
int main(){
    int n;
    while ( cin >> n ){
        int x,ret;
        cin >> ret;
        for(int i=1;i<n;++i){
            cin >> x;
            if ( x < ret ) ret = x;
        }
        cout<<ret<<endl;
    }
    return 0;
}
```

2.8 题号：10453

分段函数。

```
#include <iostream>
#include <iomanip>
#include <cmath>
using namespace std;
int main()
{
    int n;
```

```
        double x,y;
        cin>>n;
        for(int i=0;i<n;i++)
        {
            cin>>x;
            if(x<1)
                y=x;
            else if(x<10)
                y=2*x-1;
            else
                y=3*x-11;
            if(abs(y-int(y))<=1e-6)
                cout<<int(y)<<endl;
            else
                cout<<setprecision(1)<<setiosflags(ios::fixed)<<y<<endl;
        }
        return 0;
}
```

2.9 题号：10454

计算表达式的值。

```
#include<iostream>
#include<iomanip>
using namespace std;
int main()
{
    int n;
    double x,y,z,t;
    cout <<setiosflags(ios::fixed);
    cin>>n;
    while(n--)
    {
        cin>>x>>y>>z;
        t=(x+1)*(y-3)/(x+y+z);
        cout<<setprecision(4)<<t<<endl;
    }
    return 0;
```

}

2.10 题号：11603

闰年。

```
#include <iostream>
using namespace std;
int main(){
    int year;
    while ( cin >> year )
    {
        if(year%4==0 && year%100!=0) ||(year%400==0)
            cout<<"Yes";
        else
            cout<<"No";
    }
    return 0;
}
```

2.11 题号：10458

循环移位。

```
#include<iostream>
using namespace std;
void shift(int &value,int n)
{
    if(n==0) return;
    int x=1<<(sizeof(int)*8-1);
    int temp,i;
    if(n>0)
    {
        n=n%(sizeof(int)*8);
        for(i=0;i<n;i++)
        {
            temp=value&1;
            value>>=1;
            if(temp) value|=x;
        }
    }
```

```
        else
        {
            n=-n%(sizeof(int)*8);
            for(i=0;i<n;i++)
            {
                temp=value & x;
                value<<=1;
                if(temp) value|=1;
            }
        }
    }

    int main()
    {
        int count,n,m;//n为要移的位数，m为要移位的数
        cin>>count;
        while(count--)
        {
            cin>>n>>m;
            shift(m,n);
            cout<<m<<endl;
        }
        return 0;
    }
```

2.12 题号：11604

求 π。

方法一：直接用 π=acos(−1.0)计算，常利用此公式进行高精度计算。

```
#include <iostream>
#include <iomanip>
#include <cmath>
using namespace std;
double const PI = acos(-1.0);
int main(){
    int n;
    while ( cin >> n && n != -1 )
        cout<<fixed<<setprecision(n)<<PI<<endl;
```

```
    return 0;
}
```

方法二：利用题目上的公式求，得到的是近似解，由于精度原因，在 OJ 上可能通不过。掌握求解方法即可。

```
#include <iostream>
#include <iomanip>
#include <cmath>

using namespace std;

int main(){
    int k,n;
    double pi = 0;  //初始值为 0
    for(k=0;k<=10000000;k++){
        if(k%2==0)
            pi+=1.0/(2*k+1);
        else
            pi-=1.0/(2*k+1);
    }
    pi=pi*4;
    while ( cin >> n && n != -1 )
        cout<<fixed<<setprecision(n)<<pi<<endl;
    return 0;
}
```

2.13 题号：11605

求 e。

方法一：使用指数函数求。

```
#include <iostream>
#include <iomanip>
#include <cmath>
using namespace std;
double const E = exp(1);
int main(){
    int n;
    while ( cin >> n && n != -1 )
        cout<<fixed<<setprecision(n)<<E<<endl;
```

```
        return 0;
}
```
方法二：利用题目上的公式求。
```
#include <iostream>
#include <iomanip>
#include <cmath>
using namespace std;

int main(){
    int n,i;
    long double a=1;
    long double E=1.0,b;

    for (i=1;i<=150;i++)
    {
        a*=i;
        b=1.0/a;
        E+=b;

    }
    while ( cin >> n && n != -1 )
        cout<<fixed<<setprecision(n)<<E<<endl;
    return 0;
}
```

2.14 题号：11606

计算表达式的值。

```
#include<cstdio>
#include<iostream>
#include<cstring>
using namespace std;
int a,b;
char c;
int main()
{
    while(cin >> a >> c >> b){
        printf("%d%c%d=",a,c,b);
```

```
    switch(c){
        case '+':{a = a + b;break;}
        case '-':{a = a - b;break;}
        case '*':{a = a * b;break;}
        case '/':{a = a / b;break;}
        default:{a = a % b;}
    }
    printf("%d\n",a);
}
}
```

第 3 章　数组与字符串

3.1　题号：10477

杨辉三角。

```
#include<iostream>
#include<iomanip>
using namespace std;

const int Max=20;

int main()
{
    int n,i,j,k=0;
    int a[Max][Max];
    while(cin>>n)
    {
        int i,j;
        for(i=0;i<n;i++)
        {
            a[i][i]=1;
            a[i][0]=1;
        }
        for(i=2;i<n;i++)
            for(j=1;j<=i-1;j++)
```

```
            a[i][j]=a[i-1][j-1]+a[i-1][j];
        k++;
        cout<<"Case "<<k<<":"<<endl;
        for(i=0;i<n;i++)
        {
            cout<<setw((n-i-1)*3+6);
            for(j=0;j<=i-1;j++)
                cout<<a[i][j]<<setw(6);
            cout<<a[i][j]<<endl;
        }
    }
    return 0;
}
```

3.2 题号：10475

螺旋方阵。

```
#include<iostream>
using namespace std;

#define M 70
int main()
{

    int a[M][M]={0};
    int b[M]={0},count=0;
    int i,j,k,s,N,ii=0;
    while(cin>>b[count++]);
    while(--count)
    {
        N=b[ii];
        s=1;
        for(k=0;k<(N+1)/2;k++)
        {
            for(i=k,j=k;i<N-1-k;i++)    //行增加,列不变
            {
                a[i][j]=s++;
            }
```

```
            for(;j<N-1-k;j++)    //列增加,行不变
            {
                a[i][j]=s++;
            }
            for(;i>k;i--)    //行减少,列不变
            {
                a[i][j]=s++;
            }
            for(;j>k;j--)    //列减少,行不变
            {
                a[i][j]=s++;
            }
        }
        if(N%2!=0)
            a[i][j]=s;
        cout<<"n="<<b[ii++]<<endl;
        for(i=0;i<N;i++)
        {
            for(j=0;j<N-1;j++)
                cout<<a[i][j]<<" ";
                cout<<a[i][j]<<endl;
        }
        cout<<endl;
    }
    //system("pause");
    return 0;
}
```

3.3 题号：10474

在有序数列中插入数据。

```
#include<iostream>
using namespace std;

int main()
{
    int T,n,i,inumber,j;
    cin>>T;
    while(T--)
```

```
    {
        cin>>n;
        int *p= new int[n+1];
        for(i=0;i<n;i++)
            cin>>p[i];
        cin>>inumber;
        for(i=0;i<n;i++)  //寻找插入位置
        {
            if(inumber<=p[i])
                break;
        }
        for(j=n-1;j>=i;j--)//数组中从i开始的数向后移一个位置
            p[j+1]=p[j];
        p[j+1]=inumber;  //插入
        for(i=0;i<n;i++)
            cout<<p[i]<<" ";
        cout<<p[i]<<endl;
    }
    return 0;
}
```

3.4 题号：10472

矩阵乘积。

```
#include<iostream>
#include<memory>
using namespace std;

const int Max=12;
int main()
{
    int i,j,T,p,q,r,k,count=0;
    int a[Max][Max],b[Max][Max],c[Max][Max];
    cin>>T;
    while(T--)
    {
        memset(c,0,Max*Max*sizeof(int));
        count++;
```

```
        cin>>p>>q>>r;
        for(i=0;i<p;i++)   //输入数据
            for(j=0;j<q;j++)
                cin>>a[i][j];
        for(i=0;i<q;i++)   //输入数据
            for(j=0;j<r;j++)
                cin>>b[i][j];

        for(i=0;i<p;i++)
            for(j=0;j<r;j++)
                for(k=0;k<q;k++)
                    c[i][j]+=a[i][k]*b[k][j];  //矩阵乘积

        cout<<"Case "<<count<<":"<<endl;
        for(i=0;i<p;i++)    //输出矩阵
        {
            for(j=0;j<r-1;j++)
                cout<<c[i][j]<<" ";
            cout<<c[i][j]<<endl;
        }
    }
    return 0;
}
```

3.5 题号：10471

转置矩阵。

```
#include<iostream>
using namespace std;
const int Max=100;
int main()
{
    int i,j,T,m,n,k=0;
    int a[Max][Max],b[Max][Max];
    cin>>T;
    while(T--)
    {
        k++;
```

```
            cin>>m>>n;
            for(i=0;i<m;i++)   //输入数据
                for(j=0;j<n;j++)
                    cin>>a[i][j];
            for(i=0;i<m;i++)      //矩阵进行转置
                for(j=0;j<n;j++)
                    b[j][i]=a[i][j];
            cout<<"Case "<<k<<":"<<endl;
            for(i=0;i<n;i++)      //输出转置矩阵
            {
                for(j=0;j<m-1;j++)
                    cout<<b[i][j]<<" ";
                cout<<b[i][j]<<endl;
            }
        }
        return 0;
    }
```

3.6 题号：10468

按逆序显示。

```
#include<iostream>
using namespace std;
#define N 40
int main()
{
    long n,i,a[N]={0,1};
    while(cin>>n)
    {
        long i;
        for(i=2;i<n;i++)
            a[i]=a[i-2]+a[i-1];
        for(i=n-1;i>=1;i--)
            cout<<a[i]<<" ";
        cout<<a[i]<<endl;
    }
    return 0;
}
```

3.7 题号：10469

统计十进制数字字符的个数。

```cpp
#include<iostream>
using namespace std;

#define N 260
int main()
{
    int T,i,number,slen;
    char a[N];
    cin>>T;
    getchar();//请思考为何这里要用一个getchar();
    while(T--)
    {
        number=0;
        cin.getline(a,256);//这里为什么不能直接用cin
        slen=strlen(a);
        for(i=0;i<slen;i++)
            if(a[i]>='0' && a[i]<='9')
                number++;
        cout<<number<<endl;
    }
    return 0;
}
```

3.8 题号：10473

统计英文字符的个数。

```cpp
#include<iostream>
using namespace std;

#define N 258
char b[27]={'a','b','c','d','e','f','g','h','i','j','k','l',
    'm','n','o','p','q','r','s','t','u','v','w','x','y','z','\0'};

int statchar(char ch)  //求字符ch在字母表中的那个位置
{
```

```cpp
    int i;
    for(i=0;i<26;i++)
    {
        if(b[i]==ch || b[i]==ch+32)
            return i;
    }
    return 26;   //不是英文字符返回
}

int main()
{
    int T,i,j,m,c[27],cc=0;
    char a[N];
    cin>>T;
    getchar();
    while(T--)
    {
        cc++;
        for(i=0;i<26;i++)
            c[i]=0;
        cin.getline(a,256);
        m=strlen(a);
        for(i=0;i<m;i++)
        {
            j=statchar(a[i]);
            c[j]++;
        }
        cout<<"Case "<<cc<<":"<<endl;
        for(i=0;i<25;i++)
            cout<<c[i]<<" ";
        cout<<c[i]<<endl;
    }
    return 0;
}
```

3.9 题号：10476

字母大小写转换。

```cpp
#include<iostream>
using namespace std;

#define N 258
int main()
{
    int T,i,j,m;
    char a[N],b[N/2];
    cin>>T;
    getchar();
    while(T--)
    {
        j=0;
        cin.getline(a,256);
        m=strlen(a);
        for(i=1;i<m;i=i+2)
        {
            if(a[i]>='a' && a[i]<='z')  //判断是否为小写字母
            {
                b[j]=a[i]-32;  //赋值并转换为大写字母
                j++;
            }
            else
            {
                b[j]=a[i];
                j++;
            }
            b[j]='\0';
        }
        cout<<b<<endl;
    }
    return 0;
}
```

3.10 题号：10478

字符串连接。

```cpp
#include<iostream>
using namespace std;

#define N 210

int main()
{
    int T,i,j,k=0;
    char a[N],b[N];
    cin>>T;
    getchar();
    while(T--)
    {
        k++;
        i=0;
        j=0;
        cin.getline(a,100);  //读取一行字符,可以含空格
        cin.getline(b,100);
        while(a[i]!='\0')
            i++;
        while(b[j]!='\0')
            a[i++]=b[j++];
        a[i]='\0';
        cout<<"Case "<<k<<":\n"<<a<<endl;
        cin.getline(b,2);
    }
    return 0;
}
```

3.11 题号：10484

字符串解密。

```cpp
#include<iostream>
using namespace std;
```

```
#define N 1001
void encrypt(char *str)
{
    const int arr[7] ={4,9,6,2,8,7,3};
    int n,temp;
    for(n=0;n<int(strlen(str));n++)
    {
        temp=*(str+n)+arr[n%7];
        if(temp>122)
            temp=temp%122+31;
        *(str+n)=temp;
    }
}

void decryption(char *str)
{
    const int arr[7] ={4,9,6,2,8,7,3};
    int n;
    for(n=0;n<int(strlen(str));n++)
    {
        *(str+n)-=arr[n%7];
        if(*(str+n)<32)
            *(str+n)+=91;
    }
}

int main()
{
    char *str;
    int T,k=0;
    cin>>T;
    getchar();
    while(T--)
    {
        k++;
        str=new char[N];
        cin.getline(str,1000);
        cout<<"Case "<<k<<":"<<endl;
```

```
        encrypt(str);
        cout<<str<<endl<<endl;
        decryption(str);
        cout<<str<<endl;
    }
    return 0;
}
```

3.12 题号：10487

求字符串的长度。

```
#include<iostream>
using namespace std;

#define Max 260

int stringlen(char *p) //求字符串长度函数
{
    int n=0;
    while(*p!='\0')
    {
        n++;
        p++;
    }
    return n;
}


int main()
{
    int T,k=0;
    int len;
    char str[Max];
    cin>>T;
    getchar();
    while(T--)
    {
        k++;
```

```
        cin.getline(str,256);
        len=stringlen(str);
        cout<<"Case "<<k<<":"<<endl;
        cout<<len<<endl;
    }
    return 0;
}
```

3.13 题号：10367

字符串解密。

```
#include <iostream>
#include <cstdio>
#include <cstring>
using namespace std;
int main(){
    int i;
    char a[15],b[205];
    while(1){
        gets(a);
        if(strcmp(a,"ENDOFINPUT")==0) break;
        gets(b);
        gets(a);
        for(i=0;b[i]!='\0';i++){
            if(b[i]>=65&&b[i]<=90){
                b[i]-=5;
                if(b[i]<65) b[i]+=26;
            }
        }
        cout<<b<<endl;
    }
    return 0;
}
```

3.14 题号：11607

数字剥离。

```
#include <iostream>
#include <iomanip>
```

```
using namespace std;
typedef long long llt;
int main(){
    int n;
    int a[15];
    while ( cin >> n ){
        if ( 0 == n ){
            cout<<"0"<<endl;
            continue;
        }
        int k = 0;
        while(n){
            a[k++] = n % 10;
            n /= 10;
        }
        cout<<a[k-1];
        for(int i=k-2;i>=0;--i)cout<<" "<<a[i];
        cout<<endl;
    }
    return 0;
}
```

第4章 函数

4.1 题号：10452

求三个整数中最大的数。

```
#include<iostream>
using namespace std;
int threemax(int a, int b, int c)
{
    int max1;
    max1=a;  //先假定a最大
    if(a<b)  //a,b比较，求a,b中的大数
        max1=b;
    if(max1<c)  //求最大数
```

```
        max1=c;
    return max1;
}
int main()
{
    int a,b,c;
    while(cin>>a>>b>>c)   //从键盘读入 a,b,c 三个数
    {
        cout<<threemax(a,b,c)<<endl;
    }
    return 0;
}
```

4.2 题号：10454

求 n 的阶乘。

```
#include<iostream>
using namespace std;

long jc(long n)
{
    long i;
    long jc=1;
    for(i=1;i<=n;i++)
    {
        jc*=i;
    }
    return jc;
}

int main()
{
    long m,n;
    long jcc;
    cin>>m;
    while(m--)
    {
        cin>>n;
```

```
            jcc=jc(n);
            cout<<jcc<<endl;
        }
        return 0;
    }
```

4.3 题号：10459

水仙花数。

```cpp
#include<iostream>
using namespace std;

void shuixianhuashu(int n)
{
    int g,s,b;
    b=n/100;        //取出三位数的百位
    s=n/10%10;      //取出三位数的十位
    g=n%10;         //取出三位数的个位
    if(n==b*b*b+s*s*s+g*g*g)
        cout<<"YES"<<endl;
    else
        cout<<"NO"<<endl;
}

int main()
{
    int n;
    while(cin>>n)
    {
        shuixianhuashu(n);
    }
    return 0;
}
```

4.4 题号：10461

完数。

```cpp
#include<iostream>
using namespace std;
```

```
int wanshu(int n)
{
    int j,sum;  //定义数,因子
    sum=0;
    for(j=1;j<n;j++)
    {
        if(n%j==0)  //当 i 被 j 整除
        {
            //原数等于除完因子后的数
            sum+=j;  //把因子加到 sum 中
        }
    }
    if(sum==n)  //因子相加和等于原数 i 时是完数返回
        return 1;
    else
        return 0;
}

int main()
{
    int i=0,n;  //定义数,因子
    while(cin>>n)
    {
        i++;
        if(wanshu(n))  //因子相加和等于原数 i 时
            cout<<"Case "<<i<<": "<<n<<",Yes"<<endl;
        else
            cout<<"Case "<<i<<": "<<n<<",No"<<endl;
    }
    return 0;
}
```

4.5 题号：10464

素数。

```
#include<iostream>
#include<cmath>
```

```cpp
using namespace std;

int prime(int m)
{
    int i;
    for( i=2;i<=int(sqrt(m)); i++)
    {
        if ( m%i == 0 )
            return 0;
    }
    return 1;
}

int main()
{
    int T,n;
    cin>>T;
    while(T--)
    {
        cin>>n;
        if(prime(n))
            cout<<1<<endl;
        else
            cout<<0<<endl;
    }
    return 0;
}
```

4.6 题号：11601

求平均值。

```cpp
#include <iostream>
#include <iomanip>
using namespace std;
typedef long long llt;

double average(llt A[],int n)
{
```

```
        llt s = 0LL;
        for(int i=0;i<n;++i){
            s += A[i];
        }
        return (double)s/(double)n;
}
int main(){
        int n;
        llt a[1005];
        while ( cin >> n ){
            double s = 0;
            for(int i=0;i<n;++i){
                cin >> a[i];
            }
            s=average(a,n);
            cout<<fixed<<setprecision(2)<<s<<endl;
        }
        return 0;
}
```

4.7 题号：11602

求最小值。

```
#include <iostream>
using namespace std;

int minx(int a[],int n)
{
    int minvalue=a[0];
    for(int i=1;i<n;++i){
        if ( minvalue >a[i])
            minvalue = a[i];
    }
    return minvalue;
}

int main(){
    int n,a[1005];
    while ( cin >> n ){
```

```
        for(int i=0;i<n;++i){
            cin >> a[i];
        }
        cout<<minx(a,n)<<endl;
    }
    return 0;
}
```

4.8 题号：10453

分段函数。

```
#include<iostream>
#include<iomanip>
using namespace std;

double computerfunc(double x)
{
    double y;
    if(x<1)        //x<1
        y=x;
    else if(x>=10) //1<=x<10
        y=3*x-11;
    else           //x>=10
        y=2*x-1;
    return y;
}
int main()
{
    int n;
    double x,y;
    cin>>n;
    while(n--)
    {
        cin>>x;
        y=computerfunc(x);
        if ((int)y == y )
            cout<<(int)y<<endl;
        else cout<<fixed<<setprecision(1)<<y<<endl;
```

        }
        return 0;
}

4.9 题号：10454

计算表达式的值。

```
#include<iostream>
#include<iomanip>
using namespace std;

double computerfunc(double x, double y, double z)
{
    double t;
    t=(x+1)*(y-3)/(x+y+z);
    return t;
}

int main()
{
    int n;
    double x,y,z,t;
    cout <<setiosflags(ios::fixed);
    cin>>n;
    while(n--)
    {
        cin>>x>>y>>z;
        t=computerfunc(x,y,z);
        cout<<setprecision(4)<<t<<endl;
    }
        return 0;
}
```

4.10 题号：11603

闰年。

```
#include <iostream>
using namespace std;
char A[][7] = {"No\n","Yes\n"};
```

```cpp
inline int isLeap(int y){
    return 0 == y % 400
        || ( 0 == y % 4 && y % 100 );
}
int main(){
    int year;
    while ( cin >> year )
        cout << A[isLeap(year)];
    return 0;
}
```

4.11 题号：10458

循环移位。

```cpp
#include<iostream>
using namespace std;

void shift(int &value,int n)
{
    if(n==0) return;
    int x=1<<(sizeof(int)*8-1);
    int temp,i;
    if(n>0)
    {
        n=n%(sizeof(int)*8);
        for(i=0;i<n;i++)
        {
            temp=value&1;
            value>>=1;
            if(temp) value|=x;
        }
    }
    else
    {
        n=-n%(sizeof(int)*8);
        for(i=0;i<n;i++)
        {
            temp=value & x;
```

```
            value<<=1;
            if(temp) value|=1;
        }
    }
}

int main()
{
    int count,n,m;//n 为要移的位数，m 为要移位的数
    cin>>count;
    while(count--)
    {
        cin>>n>>m;
        shift(m,n);
        cout<<m<<endl;
    }
    return 0;
}
```

4.12 题号：11604

求 π。

方法一：

```
#include <iostream>
#include <iomanip>
#include <cmath>
using namespace std;
double const PI = acos(-1.0);
int main(){
    int n;
    while ( cin >> n && n != -1 )
        cout<<fixed<<setprecision(n)<<PI<<endl;
    return 0;
}
```

方法二：

```
#include <iostream>
#include <iomanip>
#include <cmath>
```

```cpp
using namespace std;
double computepi()
{
    double pi = 0;
    int k;
    for(k=0;k<=100000000;k++){
        if(k%2==0)
            pi+=1.0/(2*k+1);
        else
            pi-=1.0/(2*k+1);
    }
    pi=pi*4;
    return pi;
}

int main(){
    int n;
    double pi=computepi();
    while ( cin >> n && n != -1 )
        cout<<fixed<<setprecision(n)<<pi<<endl;
    return 0;
}
```

4.13 题号：11605

求 e。

方法一：

```cpp
#include <iostream>
#include <iomanip>
#include <cmath>
using namespace std;
double const E = exp(1);
int main(){
    int n;
    while ( cin >> n && n != -1 )
        cout<<fixed<<setprecision(n)<<E<<endl;
    return 0;
```

}

方法二：

```cpp
#include <iostream>
#include <iomanip>
#include <cmath>
using namespace std;

double computeE()
{
    int i;
    long double a=1;
    long double E=1.0,b;
    for (i=1;i<=150;i++)
    {
        a*=i;
        b=1.0/a;
        E+=b;

    }
    return E;
}

int main(){
    int n;
    double E=computeE();
    while ( cin >> n && n != -1 )
        cout<<fixed<<setprecision(n)<<E<<endl;
    return 0;
}
```

第 5 章　指针

5.1　题号：10479

交换两个数的值。

```cpp
#include<iostream>
```

```
using namespace std;

void swap(int *rx,int *ry)
{
    int temp;
    temp=*rx;
    *rx=*ry;
    *ry=temp;
}

int main()
{
    int T,a,b;
    cin>>T;
    while(T--)
    {
        cin>>a>>b;
        swap(&a,&b);
        cout<<a<<" "<<b<<endl;
    }
    return 0;
}
```

5.2 题号：10480

指针数组。

```
#include<iostream>
using namespace std;

const int M=20;
int main()
{
    char *a[M];
    int T,n,i,k=0;
    cin>>T;
    while(T--)
    {
        k++;
```

```
        cin>>n;
        getchar();
        for(i=0;i<n;i++)
        {
            a[i]=new char[M];//为指针数组动态分配存储空间
            cin.getline(a[i],20);
        }
        cout<<"Case "<<k<<":"<<endl;
        for(i=0;i<n;i++)//输出
        {
            cout<<a[i]<<endl;
        }
    }
    return 0;
}
```

5.3 题号：10481

指针访问数组元素。

```
#include<iostream>
using namespace std;

const int M=5;
int main()
{
    float a[M],*p;
    int T,i,k=0;
    cin>>T;
    while(T--)
    {
        k++;
        for(i=0;i<M;i++)
        {
            cin>>a[i];
        }
        cout<<"Case "<<k<<":"<<endl;
        for(p=a;p<a+M-1;p++)
        {
```

```
            cout<<*p<<" ";
        }
        cout<<*p<<endl;
        for(p=a+M-1;p>=a+1;p--)
        {
            cout<<*p<<" ";
        }
        cout<<*p<<endl;

        p=a;   //使指针指向数组首地址
        for(i=1;i<M;i++)  //求最高价
        {
            if(*p<a[i])
                p=a+i;
        }
        cout<<"Max:"<<*p;

        p=a;   //使指针指向数组首地址
        for(i=1;i<M;i++)  //求最低价
        {
            if(*p>a[i])
                p=a+i;
        }
        cout<<",Min:"<<*p<<endl;
    }
    return 0;
}
```

5.4 题号：10482

指针访问数组元素。

```
#include<iostream>
using namespace std;

void bubble(int a[],int n)  //冒泡排序
{
    int i,j,t,*p=a;
    for(i=0;i<n-1;i++)
```

```
    {
        for(j=0;j<n-i-1;j++)
            if(*(p+j)>*(p+j+1))
            {
                t=*(p+j);
                *(p+j)=*(p+j+1);
                *(p+j+1)=t;
            }
    }
}

int main()
{
    int *p;
    int T,n,i,k=0;
    cin>>T;
    while(T--)
    {
        k++;
        cin>>n;
        p=new int[n];
        for(i=0;i<n;i++)
        {
            cin>>p[i];
        }
        bubble(p,n);
        cout<<"Case "<<k<<":"<<endl;
        for(i=0;i<n-1;i++)
        {
            cout<<p[i]<<" ";
        }
        cout<<p[i]<<endl;
    }
    return 0;
}
```

5.5 题号：10483

字符串排序。

```cpp
#include<iostream>
using namespace std;

#define M 30
void print(char *name[],int n)  //输出
{
    int i;
    for(i=0;i<n;i++)
        cout<<name[i]<<endl;
}

void sort(char *name[],int n)  //排序
{
    int i,j,k;
    char *temp;
    for(i=0;i<n-1;i++)
    {
        k=i;
        for(j=i+1;j<n;j++)
            if(strcmp(name[k],name[j])>0)
                k=j;
        if(k!=j)
        {
            temp=name[i];
            name[i]=name[k];
            name[k]=temp;
        }
    }
}

int main()
{
    char *a[M];
    int T,n,i,k=0;
    cin>>T;
```

```
    while(T--)
    {
        k++;
        cin>>n;
        getchar();
        for(i=0;i<n;i++)
        {
            a[i]=new char[40];
            cin.getline(a[i],40);
        }
        cout<<"Case "<<k<<":"<<endl;
        print(a,n);
        cout<<endl;
        sort(a,n);
        print(a,n);
    }
    return 0;
}
```

5.6 题号：10484

字符串解密。

```
#include<iostream>
using namespace std;

#define N 1001
void encrypt(char *str)
{
    const int arr[7] ={4,9,6,2,8,7,3};
    int n,temp;
    for(n=0;n<int(strlen(str));n++)
    {
        temp=*(str+n)+arr[n%7];
        if(temp>122)
            temp=temp%122+31;
        *(str+n)=temp;
    }
}
```

```cpp
void decryption(char *str)
{
    const int arr[7] ={4,9,6,2,8,7,3};
    int n;
    for(n=0;n<int(strlen(str));n++)
    {
        *(str+n)-=arr[n%7];
        if(*(str+n)<32)
            *(str+n)+=91;
    }
}

int main()
{
    char *str;
    int T,k=0;
    cin>>T;
    getchar();
    while(T--)
    {
        k++;
        str=new char[N];
        cin.getline(str,1000);
        cout<<"Case "<<k<<":"<<endl;
        encrypt(str);
        cout<<str<<endl<<endl;
        decryption(str);
        cout<<str<<endl;
    }
    return 0;
}
```

5.7 题号：10485

二维数组的使用。

```cpp
#include<iostream>
using namespace std;
```

```c
#define M 3
#define N 4
float minimum(float *p,int n)  //求最低成绩
{
    float min=0;
    float *p_end;
    p_end=p+n-1;
    min=*p;
    for(++p;p<p_end;p++)
    {
        if(min>*p)
            min=*p;
    }
    return min;
}
float maximum(float *p,int n)  //求最高成绩
{
    float max=0;
    float *p_end;
    p_end=p+n-1;
    max=*p;
    for(++p;p<p_end;p++)
    {
        if(max<*p)
            max=*p;
    }
    return max;
}
void average(float (*p)[N],float b[],int m,int n)/*求每个学生的平均成绩*/
{
    int i,j;
    float sum=0;
    for(i=0;i<m;i++)
    {
        for(j=0;j<n;j++)
            sum+=*(*(p+i)+j);
        b[i]=sum/n;
```

```cpp
            sum=0;
        }
    }
    void printArray(float (*p)[N],float b[],int m,int n)  //输出
    {
        int i,j;
        cout<<"No  "<<"kc1  "<<"kc2  "<<"kc3  "<<"kc4  "
            <<"average_score"<<endl;
        for(i=0;i<m;i++)
        {
            cout<<i+1<<"    ";
            for(j=0;j<n;j++)
                cout<<*(*(p+i)+j)<<"    ";
            cout<<b[i]<<endl;
        }
    }

    int main()
    {
        float a[M][N];
        float b[M];
        int T,i,j,k=0;
        cin>>T;
        while(T--)
        {
            k++;
            for(i=0;i<M;i++)
                for(j=0;j<N;j++)
                    cin>>a[i][j];
            cout<<"Case "<<k<<":"<<endl;
            cout<<"Min score:"<<minimum(*a,M*N)<<endl;
            cout<<"Max score:"<<maximum(*a,M*N)<<endl;
            average(a,b,M,N);
            printArray(a,b,M,N);
        }
        return 0;
    }
```

5.8 题号：10486

函数指针数组。

```cpp
#include<iostream>
using namespace std;

#define M 3
#define N 4
void mininum(float (*p)[N],float *min,int m,int n)
{
    int i,j;
    *min=*(*(p+0)+0);
    for(i=0;i<m;i++)
    {
        for(j=0;j<n;j++)
            if(*min>*(*(p+i)+j))
                *min=*(*(p+i)+j);
    }
    cout<<"Min score:"<<*min<<endl;
}
void maxinum(float (*p)[N],float *max,int m,int n)
{
    int i,j;
    *max=*(*(p+0)+0);
    for(i=0;i<m;i++)
    {
        for(j=0;j<n;j++)
            if(*max<*(*(p+i)+j))
                *max=*(*(p+i)+j);
    }
    cout<<"Max score:"<<*max<<endl;
}
void average(float (*p)[N],float *b,int m,int n)
{
    int i,j;
    float sum=0;
    cout<<"Average_score:\n";
    for(i=0;i<m;i++)
```

```cpp
    {
        for(j=0;j<n;j++)
            sum+=*(*(p+i)+j);
        b[i]=sum/n;
        sum=0;
        cout<<i+1<<"   "<<b[i]<<endl;
    }
}
void printArray(float (*p)[N],float *b,int m,int n)
{
    int i,j;
    cout<<"No   "<<"kc1   "<<"kc2   "<<"kc3   "<<"kc4   "<<endl;
    for(i=0;i<m;i++)
    {
        cout<<i+1<<"   ";
        for(j=0;j<n;j++)
            cout<<*(*(p+i)+j)<<"   ";
        cout<<endl;
    }
}

int main()
{
    float a[M][N];
    float b[M];
    int T,i,j,k=0;
    int choice,n;
    void (*f[4])(float (*)[4],float *,int,int)={printArray,
        mininum,maxinum,average};
    cin>>T;
    while(T--)
    {
        k++;
        for(i=0;i<M;i++)
            for(j=0;j<N;j++)
                cin>>a[i][j];
        cout<<"Case "<<k<<":"<<endl;
        cout<<"Enter a choice:\n"
```

```
            <<"0  Print the array of grades\n"
            <<"1  Find the mininum grade\n"
            <<"2  Find the maxinum grade\n"
            <<"3  Print the average on all tests for each student\n"
            <<"4  End program\n";
        cin>>n;
        while(n--)
        {
            cin>>choice;
            if(choice==4)
                break;
            (*f[choice])(a,b,M,N);
        }
    }
    return 0;
}
```

5.9 自动产生文章。

```
#include<iostream>
using namespace std;

#define M 5
#define N 30
#define K 20

int main()
{
    char *article[M]={"the","a","one","some","any"};
    char *noun[M]={"boy","girl","dog","town","car"};
    char *verb[M]={"drove","jumped","ran","walked","skipped"};
    char *preposition[M]={"to","from","over","under","on"};
    char sentence[K][N]={""};
    int i,m,T,n,count=0;
    cin>>T;
    while(T--)
    {
        count++;
        cin>>n;
```

```
        for(i=0;i<n;i++)    //产生n个句子
        {
            m=rand()%5;  //产生随机数0~4
            strcat(sentence[i],article[m]);  //选择冠词并连接到句子中
            strcat(sentence[i]," ");    //加入一空格
            m=rand()%5;  //产生随机数0~4
            strcat(sentence[i],noun[m]);  //选择名词并连接到句子中
            strcat(sentence[i]," ");   //加入一空格
            m=rand()%5;  //产生随机数0~4
            strcat(sentence[i],verb[m]);  //选择动词并连接到句子中
            strcat(sentence[i]," ");   //加入一空格
            m=rand()%5;  //产生随机数0~4
            strcat(sentence[i],preposition[m]);  //选择介词并连接到句子中
            strcat(sentence[i],".");   //句子结束,加入圆点
            if(sentence[i][0]>='a' && sentence[i][0]<='z')
                sentence[i][0]-=32;  //将句子的第一个字符转换成大写
        }
        cout<<"Case "<<count<<":"<<endl;
        for(i=0;i<n;i++)  //输出文章
        {
            cout<<sentence[i];
            memset(sentence[i],0,sizeof(sentence[i]));  //清空字符数组内容
        }
        cout<<endl;

    }
    return 0;
}
```

5.10 题号：10487

求字符串的长度。

```
#include<iostream>
using namespace std;

#define Max 260

int stringlen(char *p)  //求字符串长度函数
```

```
{
    int n=0;
    while(*p!='\0')
    {
        n++;
        p++;
    }
    return n;
}


int main()
{
    int T,k=0;
    int len;
    char str[Max];
    cin>>T;
    getchar();
    while(T--)
    {
        k++;
        cin.getline(str,256);
        len=stringlen(str);
        cout<<"Case "<<k<<":"<<endl;
        cout<<len<<endl;
    }
    return 0;
}
```

第6章 结构体与共用体

6.1 题号：10488

计算某日在本年中是第几天。

```
#include<iostream>
using namespace std;
```

```cpp
//day_tab 二维数组存放各月天数,第一行对应非闰年,第二行对应闰年
int day_tab[2][12]={{31,28,31,30,31,30,31,31,30,31,30,31},
{31,29,31,30, 31,30,31,31,30,31,30,31}};
struct date
{
    int year;
    int month;
    int day;
};
int leap(int year)   //闰年函数
{
    if(year%4==0&&year%100!=0||year%400==0)  //是闰年
        return 1;
    else  // 不是闰年
        return 0;
}
int dton(struct date rq)  //将日期转换成是本年的第多少天
{
    int m,days=0;
    for(m=0;m<rq.month-1;m++)
        if(leap(rq.year))  //是闰年
            days+=day_tab[1][m];
        else //不是闰年
            days+=day_tab[0][m];
        days+=rq.day;
        return days;
}

int main()
{
    struct date d;
    int day_number;
    while(cin>>d.year && cin>>d.month && cin>>d.day)
    {
        day_number=dton(d);
        cout<<day_number<<endl;
    }
```

        return 0;
}

6.2 题号：10489

复数。

```
#include<iostream>
using namespace std;

struct complex
{
    int real; //复数的实部
    int im; //复数的虚部
};
struct complex cadd(struct complex c1,struct complex c2) //求复数的和
{
    struct complex result;
    result.real=c1.real+c2.real;
    result.im=c1.im+c2.im;
    return result;
}
struct complex cmult(struct complex c1,struct complex c2) //求复数的积
{
    struct complex result;
    result.real=c1.real*c2.real-c1.im*c2.im;
    result.im=c1.real*c2.im+c1.im*c2.real;
    return result;
}
void puts(struct complex c) //输出复数
{
    cout<<c.real<<"+("<<c.im<<"i)"<<endl;
}

int main()
{
    struct complex c1,c2,c3;
    while(cin>>c1.real && cin>>c1.im && cin>>c2.real && cin>>c2.im)
    {
```

```
            c3=cadd(c1,c2);
            puts(c3);
            c3=cmult(c1,c2);
            puts(c3);
        }
        return 0;
    }
```

6.3 题号：10490

结构体排序。

```
#include<iostream>
#include<iomanip>
using namespace std;

#define M 20
struct bookinfor
{
    char name[M];
    float price;
    char author[M];
    char press[M];
}book[30];

void SelectSort(struct bookinfor b[],int n)  //对字符串进行选择排序算法
{
    int i,j,k;
    struct bookinfor temp;
    for(i=1;i<n;i++)
    {
        //进行 M-1 次选择和交换
        k=i-1;  //给 k 赋初值
        for(j=i;j<n;j++)  //选出当前区间内的最小值a[k]
            if(strcmp(b[j].name,b[k].name)<0)
                k=j;      //进行字符串比较
        temp=b[i-1];
        b[i-1]=b[k];
        b[k]=temp;
```

```cpp
        }
    }

    int main()
    {
        int i,n;
        while(cin>>n)
        {
            for(i=0;i<n;i++)
                cin>>book[i].name>>book[i].price>>book[i].
                    author>>book[i].press;
            SelectSort(book,n);
            for(i=0;i<n;i++)
            {
                cout.setf(ios::left);
                cout<<setw(20)<<book[i].name<<setw(20)<<book[i].price
                    <<setw(20)<<book[i].author<<setw(20)<<book[i].
                        press<<endl;
            }
            cout<<endl;
        }
        return 0;
    }
```

6.4 题号：10491

报数。

```cpp
#include<iostream>
using namespace std;
#define M 200
struct person
{
    int number;
    int nextp;
};
person link[M+1];

int main()
```

```cpp
{
    int i,j,count,h,T,N;
    cin>>T;
    for(j=1;j<=T;j++)
    {
        cin>>N;
        for(i=1;i<=N;i++)//初始化
        {
            if(i==N)
                link[i].nextp=1;
            else
                link[i].nextp=i+1;
            link[i].number=i;
        }
        count=0;
        h=N;
        cout<<"Sequence that persons leave the circle:"<<endl;
        while(count<N-1)
        {
            i=0;
            while(i!=3)//报数
            {
                h=link[h].nextp;
                if(link[h].number)
                    i++;
            }
            cout<<link[h].number<<" ";
            link[h].number=0;
            count++;
        }
        cout<<"\nThe last one is:";
        for(i=1;i<=N;i++)
            if(link[i].number)
                cout<<link[i].number<<endl;
    }
    return 0;
}
```

6.5 题号：10492

枚举类型。

```
#include<iostream>
using namespace std;

int main()
{
    enum color{red,yellow,blue,green,white,black};
    enum color col;
    int i;
    while(cin>>i)
    {
        col=(enum color)i;
        switch(col)
        {
        case red:cout<<"red";break;
        case yellow:cout<<"yellow";break;
        case blue:cout<<"blue";break;
        case green:cout<<"green";break;
        case white:cout<<"white";break;
        case black:cout<<"black";break;
        }
        cout<<endl;
    }
    return 0;
}
```

第7章 类与对象及封装性

7.1 运行结果。

```
Meow.
frisky is a cat who is5years old.
Meow.
```

7.2 题号：10495

用类实现复数的加法和乘法。

```cpp
#include<iostream>
using namespace std;

class complex
{
    float real;
    float im;
public:
    complex(float m=0,float n=0);
    void set_c(float m,float n);
    void cadd(complex c1,complex c2);
    void cmult(complex c1,complex c2);
    void put();
};
complex::complex(float m,float n)
{
    real=m;
    im=n;
}
void complex::set_c(float m,float n)
{
    real=m;
    im=n;
}
void complex::cadd(complex c1,complex c2)
{
    real=c1.real+c2.real;
    im=c1.im+c2.im;
}
void complex::cmult(complex c1,complex c2)
{
    real=c1.real*c2.real-c1.im*c2.im;
    im=c1.real*c2.im+c1.im*c2.real;
}
void complex::put()
```

```
{
    cout<<real<<"+("<<im<<"i)"<<endl;
}

int main( )
{
    complex c1,c2,c3;
    float   r1,r2,i1,i2;
    while(cin>>r1 && cin>>i1 && cin>>r2 && cin>>i2)
    {
        c1.set_c(r1,i1);
        c2.set_c(r2,i2);
        c3.cadd(c1,c2);
        c3.put();
        c3.cmult(c1,c2);
        c3.put();
    }
    return 0;
}
```

7.3 题号：10496

定义日期类。

```
#include<iostream>
using namespace std;

int day_tab[12]={31,28,31,30,31,30,31,31,30,31,30,31};
//day_tab二维数组存放各月天数，第一行对应非闰年，第二行对应闰年
class Date
{
    int year,month,day;
    int leap(int);//判断指定的年份是否为闰年
public:
    Date(){}
    Date(int d,int m,int y);
    void setdate(int d,int m,int y);
    void addoneday();//返回一日期加一天得到的日期
    void outputdate();
```

```cpp
};
int Date::leap(int year)
{
    if(year%4==0&&year%100!=0||year%400==0) //是闰年
        return 1;
    else //不是闰年
        return 0;
}
Date::Date(int d,int m,int y)
{
    day=d;
    month=m;
    year=y;
}
void Date::addoneday()//加一天
{
    if(month==12&&day==31)//一年的年底
    {
        year=year+1; //年加1
        month=1; //月为下一年的1月
        day=1; //日为1日
    }
    else if(leap(year)&&month==2&&day==29)//是闰年的第二月的月底
    {
        month=month+1; //3月
        day=1;    //1日
    }
    else if(day<day_tab[month-1]||(leap(year)&&month==2&&day==28))
        //日前的天数小于本月的最后一天，或是闰年的2月28日
        day=day+1; //天数加1
    else //每个月的月底
    {
        month=month+1;
        day=1;
    }
}
void Date::setdate(int d,int m,int y)
{
```

```
        day=d;
        month=m;
        year=y;
}
void Date::outputdate()
{
        cout<<day<<"/"<<month<<"/"<<year<<endl;
}


int main( )
{
        Date d1,d2;
        int  d,m,y;
        while(cin>>d && cin>>m && cin>>y)
        {
            d1.setdate(d,m,y);
            d1.outputdate();
            d1.addoneday();
            d1.outputdate();
        }
        return 0;
}
```

第 8 章　类的深入

8.1　运行结果。

Normal constructor
Copy constructor
10Destructed
Destructed

8.2　题号：10497

用类和友元函数来实现复数的加法和乘法。

```
#include<iostream>
```

```cpp
using namespace std;

class complex
{
    float real;
    float im;
public:
    complex(float m=0,float n=0);
    void set_c(float m,float n);
    friend complex cadd(complex c1,complex c2);
    friend complex cmult(complex c1,complex c2);
    void put();
};
complex::complex(float m,float n)
{
    real=m;
    im=n;
}
void complex::set_c(float m,float n)
{
    real=m;
    im=n;
}
complex cadd(complex c1,complex c2)
{
    complex result;
    result.real=c1.real+c2.real;
    result.im=c1.im+c2.im;
    return result;
}
complex cmult(complex c1,complex c2)
{
    complex result;
    result.real=c1.real*c2.real-c1.im*c2.im;
    result.im=c1.real*c2.im+c1.im*c2.real;
    return result;
}
void complex::put()
```

```
    {
        cout<<real<<"+("<<im<<"i)"<<endl;
    }

int main( )
{
    complex c1,c2,c3;
    float   r1,r2,i1,i2;
    while(cin>>r1 && cin>>i1 && cin>>r2 && cin>>i2)
    {
        c1.set_c(r1,i1);
        c2.set_c(r2,i2);
        c3=cadd(c1,c2);
        c3.put();
        c3=cmult(c1,c2);
        c3.put();
    }
    return 0;
}
```

8.3 题号：10498

求两个日期之间相差的天数。

```
#include<iostream>
using namespace std;

class Date
{
    int year;
    int month;
    int day;
public:
    Date(int y=0,int m=0,int d=0)
    {
        year=y;month=m;day=d;
    }
    void setdate(int y,int m,int d)
    {
```

```cpp
        year=y;
        month=m;
        day=d;
    }
    void disp()
    {
        cout<<year<<"/"<<month<<"/"<<day;
    }
    friend int count_day(Date &d,int);
    friend int leap(int year);
    friend int subs(Date &d1,Date &d2);
};
int count_day(Date &d,int flag)
{
    static int day_tab[2][12]={{31,28,31,30,31,30,31,31,30,31,30,31},
    {31,29,31,30,31,30,31,31,30,31,30,31}};
    //使用二维数组存放各月天数,第一行对应非闰年,第二行对应闰年
    int p,i,s;
    if(leap(d.year))
        p=1;
    else p=0;
    if(flag)
    {
        s=d.day;
        for(i=1;i<d.month;i++)
            s+=day_tab[p][i-1];
    }
    else
    {
        s=day_tab[p][d.month-1]-d.day;
        for(i=d.month+1; i<=12; i++)
            s+=day_tab[p][i-1];
    }
    return s;
}
int leap(int year)
{
    if(year%4==0&&year%100!=0||year%400==0)//是闰年
```

```
            return 1;
        else  //不是闰年
            return 0;
}
int subs(Date &d1,Date &d2)//两日期相减
{
    int days,day1,day2,y;
    if(d1.year<d2.year)
    {
        days=count_day(d1,0);
        for(y=d1.year+1; y<d2.year;y++)
            if(leap(y))
                days+=366;
            else
                days+=365;
        days+=count_day(d2,1);
    }
    else if(d1.year==d2.year)
    {
        day1=count_day(d1,1);
        day2=count_day(d2,1);
        days=day2-day1;
    }
    else
        days=-1;
    return days;
}

int main( )
{
    Date d1,d2;
    int  day1,m1,y1,day2,m2,y2,ds;
    while(cin>>y1 && cin>>m1 && cin>>day1 && cin>>y2 && cin>>m2 && cin>>day2)
    {
        d1.setdate(y1,m1,day1);
        d2.setdate(y2,m2,day2);
        ds=subs(d1,d2);
        d2.disp();
```

```cpp
        cout<<"-";
        d1.disp();
        cout<<"="<<ds<<endl;
    }
    return 0;
}
```

8.4 题号：11608

对象传入函数。

```cpp
#include <iostream>
#include <cstdio>
#include <cmath>
using namespace std;

class tri{
public:
    double a,b,c;
tri(double x, double y,double z)
{a = x; b = y; c = z;
printf("Constructing\n");}
    ~tri(){printf("Destructed\n");};
};
double Area(tri T){
    double s = (T.a+T.b+T.c)/2;
    double area = sqrt(s *(s-T.a)*(s-T.b)*(s-T.c));
    return area;
}

int main()
{
    double x,y,z;
    while(scanf("%lf%lf%lf", &x,&y,&z)!=EOF){
        tri T(x,y,z);
        double area = Area(T);
        printf("%.2lf\n", area);
    }
    return 0;
```

}

8.5 题号：11609

函数返回对象。

```
#include <cstdio>
#include <iostream>
#include <cmath>
using namespace std;
#define PI acos(-1.0)
class point{
public:
    double x;
    double y;
    point(double x,double y){this->x = x;this->y = y;}
};
point tran(point tep,double a){
    return point(tep.x*cos(a) - tep.y*sin(a) ,tep.x*sin(a) +
        tep.y*cos(a));
}
int main()
{
    double x,y,a;
    while(scanf("%lf%lf%lf",&x,&y,&a)!=EOF){
        point p(x,y);
        point p1 = tran(p,a/180*PI);
        printf("%.2lf %.2lf\n",p1.x,p1.y);
    }
    return 0;
}
```

8.6 题号：11610

综合题目。

```
#include <cstdio>
#include <cmath>
#include <iostream>
using namespace std;
```

```cpp
class point{
    double x,y;
public:
    point(double x,double y){this->x = x; this->y = y;}
    point(const point&ob){x = ob.x; y = ob.y; printf(
        "copy constructor\n");}
    friend double area(point a,point b,point c);
};
double area(point a,point b,point c){
    double e1 = sqrt((a.x - b.x)*(a.x - b.x) + (a.y -b.y)*(a.y - b.y));
    double e2 = sqrt((a.x - c.x)*(a.x - c.x) + (a.y -c.y)*(a.y - c.y));
    double e3 = sqrt((b.x - c.x)*(b.x - c.x) + (b.y -c.y)*(b.y - c.y));
    double s = (e1+e2+e3)/2.0;
    double area1 = sqrt(s *(s-e1)*(s-e2)*(s-e3));
    return area1;
}
int main(){
    double x1,y1,x2,y2,x3,y3;
    while(scanf("%lf%lf%lf%lf%lf%lf",&x1,&y1,&x2,&y2,&x3,&y3) != EOF){
        point A(x1,y1),B(x2,y2),C(x3,y3);
        printf("%.2lf\n",area(A,B,C));
    }
    return 0;
}
```

第9章 运算符重载

9.1 题号：10499

复数类的加法与减法。

方法一：

```cpp
#include<iostream.h>
//using namespace std;

class complex
```

```cpp
{
    float real;
    float im;
public:
    complex(){real=im=0;}
    complex(float m,float n);
    void set_c(float m,float n);
    complex operator+(complex c);
    friend complex operator+(complex c,float i);
    friend complex operator+(float i,complex c);
    void put();
};
complex::complex(float m,float n)
{
    real=m;
    im=n;
}
void complex::set_c(float m,float n)
{
    real=m;
    im=n;
}
complex complex::operator+(complex c)
{
    complex result;
    result.real=real+c.real;
    result.im=im+c.im;
    return result;
}
complex operator+(complex c,float i)
{
    complex result;
    result.real=c.real+i;
    result.im=c.im;
    return result;
}
complex operator+(float i,complex c)
{
```

```cpp
    complex result;
    result.real=i+c.real;
    result.im=c.im;
    return result;
}
void complex::put()
{
    cout<<real<<"+("<<im<<"i)"<<endl;
}

int main( )
{
    complex a,b,c;
    float r1,r2,i1,i2,m,n;
    while(cin>>r1 && cin>>i1 && cin>>r2 && cin>>i2)
    {
        cin>>m>>n;
        a.set_c(r1,i1);
        b.set_c(r2,i2);
        c=a+b;
        c.put();
        c=m+a;
        c.put();
        c=b+n;
        c.put();
    }
    return 0;
}
```

方法二:
```cpp
#include<iostream.h>
//using namespace std;

class Complex
{
public:
    Complex(float r=0, float v=0):real(r),im(v){}
    void set_c(float m,float n);
    friend Complex operator+(Complex a, Complex b);
```

```cpp
    friend ostream& operator<<(ostream& out, Complex& a);
private:
    float real;
    float im;
};
void Complex::set_c(float m,float n)
{
    real=m;
    im=n;
}
ostream& operator<<(ostream& out, Complex& a)
{
    return out<<a.real<<"+("<<a.im<<"i)"<<endl;
}
Complex operator+(Complex a, Complex b)
{
    return Complex(a.real+b.real,a.im+b.im);
}

int main( )
{
    Complex a,b,c;
    float  r1,r2,i1,i2,m,n;
    while(cin>>r1 && cin>>i1 && cin>>r2 && cin>>i2)
    {
        cin>>m>>n;
        a.set_c(r1,i1);
        b.set_c(r2,i2);
        c=a+b;
        cout<<c;
        c=m+a;
        cout<<c;
        c=b+n;
        cout<<c;
    }
    return 0;
}
```

9.2 题号：10500

时间类。

方法一：

```cpp
#include<iostream>
using namespace std;

class Time
{
    int hour;
    int minute;
    int second;
public:
    Time(int h=0,int m=0,int s=0);
    void SetTime(int h,int m,int s);
    int GetHour()
    {
        return hour;
    }
    int GetMinute()
    {
        return minute;
    }
    int GetSecond()
    {
        return second;
    }
    Time operator+(Time t);//返回一时间加一时间得到的时间
    Time operator-(Time t);//返回一时间减一时间得到的时间
    void OutputDate();
};
Time::Time(int h,int m,int s)
{
    hour=h;
    minute=m;
    second=s;
}
void Time::SetTime(int h,int m,int s)
```

```cpp
{
    hour=h;
    minute=m;
    second=s;
}
Time Time::operator+(Time t)
{
    Time temp;
    temp.hour=hour+t.hour;
    temp.minute=minute+t.minute;
    temp.second=second+t.second;
    if(temp.second>=60)  //相加后秒大于等于60
    {
        temp.minute+=temp.second/60;  //60秒转化为1分钟
        temp.second%=60;
    }
    if(temp.minute>=60)  //分大于等于60
    {
        temp.hour+=temp.minute/60;  //60分钟转化为1小时
        temp.minute%=60;
    }
    if(temp.hour>=24)  //时间大于等于24小时
    {
        temp.hour%=24;
    }
    return temp;
}
Time Time::operator-(Time t)
{
    Time temp;
    temp.hour=hour-t.hour;
    temp.minute=minute-t.minute;
    temp.second=second-t.second;
    if(temp.second<0)  //减后秒小于0
    {
        temp.minute--;//从分钟借1分钟到秒
        temp.second+=60;
    }
```

```cpp
        if(temp.minute<0)  //分钟小于0
        {
            temp.hour--;//从小时借1小时到分钟
            temp.minute+=60;
        }
        if(temp.hour<0)   //小时小于0
        {
            temp.hour+=24;
        }
    return temp;
}
void Time::OutputDate()
{
    cout<<hour<<":"<<minute<<":"<<second;
}

int main( )
{
    Time a,b,c;
    int h1,m1,s1,h2,m2,s2;
    while(cin>>h1 && cin>>m1 && cin>>s1 && cin>>h2 && cin>>m2 && cin>>s2)
    {
        a.SetTime(h1,m1,s1);
        b.SetTime(h2,m2,s2);
        a.OutputDate();
        cout<<"+";
        b.OutputDate();
        c=a+b;
        cout<<"=";
        c.OutputDate();
        cout<<endl;
        a.OutputDate();
        cout<<"-";
        b.OutputDate();
        cout<<"=";
        c=a-b;
        c.OutputDate();
        cout<<endl;
```

        }
        return 0;
}
方法二：
```
#include<iostream>
using namespace std;

class Time
{
    int hour;
    int minute;
    int second;
public:
    Time(int h=0,int m=0,int s=0);
    void SetTime(int h,int m,int s);
    int GetHour()
    {
        return hour;
    }
    int GetMinute()
    {
        return minute;
    }
    int GetSecond()
    {
        return second;
    }
    Time operator+(Time t);//返回一时间加一时间得到的时间
    Time operator-(Time t);//返回一时间减一时间得到的时间
    void OutputDate();
};
Time::Time(int h,int m,int s)
{
    hour=h;
    minute=m;
    second=s;
}
void Time::SetTime(int h,int m,int s)
```

```cpp
{
    hour=h;
    minute=m;
    second=s;
}
Time Time::operator+(Time t)
{
    int temp=(hour+t.hour)*3600+(minute+t.minute)*60+(second+t.second);
    //将时间转换为秒
    return Time(temp/3600%24,temp%3600/60,temp%60);  //返回转换后的时间
}
Time Time::operator-(Time t)
{
    int temp=(hour-t.hour)*3600+(minute-t.minute)*60+(second-t.second);
//将时间转换为秒
    if(temp<0)
        temp=24*3600-(-temp)%(24*3600);
    return Time(temp/3600%24,temp%3600/60,temp%60);
}
void Time::OutputDate()
{
    cout<<hour<<":"<<minute<<":"<<second;
}

int main( )
{
    Time a,b,c;
    int h1,m1,s1,h2,m2,s2;
    while(cin>>h1 && cin>>m1 && cin>>s1 && cin>>h2 && cin>>m2 && cin>>s2)
    {
        a.SetTime(h1,m1,s1);
        b.SetTime(h2,m2,s2);
        a.OutputDate();
        cout<<"+";
        b.OutputDate();
        c=a+b;
        cout<<"=";
        c.OutputDate();
```

```
            cout<<endl;
            a.OutputDate();
            cout<<"-";
            b.OutputDate();
            cout<<"=";
            c=a-b;
            c.OutputDate();
            cout<<endl;
        }
        return 0;
    }
```

9.3 题号：10502

分数类。

```
#include<iostream.h>
#include<stdlib.h>
//using namespace std;

class RationalNumber
{
    int numerator; //分子
    int denominator; //分母
    int GetMax(int n, int m);//最大公约数
public:
    RationalNumber(int nt=0, int dt=1);
    RationalNumber(const RationalNumber& r);
    void Convert();
    void show()
    {
        if (abs(numerator)==abs(denominator))
            cout<<numerator/denominator;
        else
            cout<<numerator<<"/"<<denominator;
    }
    bool operator > ( const RationalNumber& r);
    bool operator < (const RationalNumber& r);
    bool operator == (const RationalNumber& r);
```

```cpp
        friend RationalNumber operator+ (const RationalNumber& a,
            const RationalNumber& b);
};

RationalNumber::RationalNumber(int nt, int dt)
{
    if(dt == 0)
    {
        cout<<"denominator equal zero";
        exit(1);
    }
    else if(dt<0)
    {
        nt = -nt;
        dt = -dt;
    }
    numerator=nt;
    denominator=dt;
    if(numerator==0)
        denominator = 1;
    else
        Convert();
}
RationalNumber::RationalNumber(const RationalNumber& r)
{
    numerator = r.numerator;
    denominator = r.denominator;
}
int RationalNumber::GetMax(int n, int m)//最大公约数
{
    int t,r;
    if(m>n)
    {
        t=m;
        m=n;
        n=t;
    }
    while(m!=0)//求两个数n,m的最大公约数
```

```cpp
        {
            r=n%m;
            n=m;
            m=r;
        }
        return n;
}
void RationalNumber::Convert()
{
    int g= GetMax(abs(numerator),abs(denominator));
    numerator/=g;
    denominator/=g;
}
bool RationalNumber::operator > ( const RationalNumber& r)
{
    return (numerator * r.denominator > denominator * r.numerator);
}
bool RationalNumber::operator < (const RationalNumber& r)
{
    return (numerator * r.denominator <denominator * r.numerator);
}
bool RationalNumber::operator == (const RationalNumber& r)
{
    return (numerator * r.denominator ==denominator * r.numerator);
}
RationalNumber operator+ (const RationalNumber& a,
    const RationalNumber& b)
{
    return RationalNumber(a.numerator * b.denominator +
        b.numerator * a.denominator, a.denominator * b.denominator);
}

int main()
{
    int T,n1,d1,n2,d2;
    cin>>T;
    while(T--)
    {
```

```
        cin>>n1>>d1>>n2>>d2;
        RationalNumber a(n1,d1),b(n2,d2),c;
        bool bflag;
        a.show();
        cout<<"+";
        b.show();
        cout<<"=";
        c=a+b;
        c.show();
        cout<<endl;

        a.show();
        cout<<">";
        b.show();
        cout<<"=";
        bflag=a>b;
        cout<<bflag<<endl;

        a.show();
        cout<<"<";
        b.show();
        cout<<"=";
        bflag=a<b;
        cout<<bflag<<endl;

        a.show();
        cout<<"==";
        b.show();
        cout<<"=";
        bflag=a==b;
        cout<<bflag<<endl;
    }
        return 0;
}
```

第 10 章 继承性

10.1 继承的访问权限问题。

```
class derived1:private base
class derived1:public base    //修改为public即可
```

源程序编译会出现如下错误：

错误 1   error C2248: "base::i": 无法访问 无法访问的成员 (在 "base" 类中声明)   d:\test\test\test\test.cpp   38

错误 2   error C2248: "base::j": 无法访问 无法访问的成员 (在 "base" 类中声明)   d:\test\test\test\test.cpp   38

错误 3   error C2247: "base::set" 不可访问，因为 "derived1" 使用 "private" 从 "base" 继承   d:\test\test\test\test.cpp   45

错误 4   error C2247: "base::show" 不可访问，因为 "derived1" 使用 "private" 从 "base" 继承   d:\test\test\test\test.cpp   46

错误 5   error C2247: "base::set" 不可访问，因为 "derived1" 使用 "private" 从 "base" 继承   d:\test\test\test\test.cpp   47

错误 6   error C2247: "base::show" 不可访问，因为 "derived1" 使用 "private" 从 "base" 继承   d:\test\test\test\test.cpp   48

10.2 题号：10503

继承。

```
#include<iostream>
using namespace std;

class Building
{
public:
    Building(int f,int r,double ft)
    {
        floors=f;
        rooms=r;
        footage=ft;
    }
    void show()
    {
```

```cpp
            cout<<"Floors:"<<floors<<endl;
            cout<<"Rooms:"<<rooms<<endl;
            cout<<"Total area:"<<footage<<endl;
        }
    protected:
        int floors;
        int rooms;
        double footage;
};
class Housing:public Building
{
public:
    Housing(int f,int r,double ft,int bd,int bth):Building(f,r,ft)
    {
        bedrooms=bd;
        bathrooms=bth;
    }
    void show()
    {
        cout<<"HOUSING:"<<endl;
        Building::show();
        cout<<"Bedrooms:"<<bedrooms<<endl;
        cout<<"Bathrooms:"<<bathrooms<<endl;
    }
private:
    int bedrooms;
    int bathrooms;
};
class Office:public Building
{
public:
    Office(int f,int r,double ft,int ex,int ph):Building(f,r,ft)
    {
        phones=ph;
        extinguishers=ex;
    }
    void show()
    {
```

```cpp
        cout<<"OFFICING:"<<endl;
        Building::show();
        cout<<"Extinguishers:"<<extinguishers<<endl;
        cout<<"Phones:"<<phones<<endl;
    }
private:
    int phones;
    int extinguishers;
};

int main()
{
    int T,f,r,bd,bth,ph,ex;
    double ft;
    cin>>T;
    while(T--)
    {
        cin>>f>>r>>ft>>bd>>bth;
        Housing hob(f,r,ft,bd,bth);
        cin>>f>>r>>ft>>ex>>ph;
        Office oob(f,r,ft,ex,ph);
        hob.show();
        oob.show();
    }
        return 0;
}
```

10.3 题号：11611

类的继承。

```cpp
#include<iostream>
#include<cstring>
#include<cstdio>
using namespace std;
class Iphone5
{
    public :
        char Color[11];
```

```cpp
        int Prize;
        void Set_Color(char c[11])
        {
            strcpy(Color,c);
        }
        void Set_Prize(int x)
        {
            Prize=x;
        }
        void Show()
        {
            cout<<"Iphone5 { Color : "<<Color<<" \\ Prize : "<<Prize<<" }"<<endl;
        }
};

class Iphone5S: public Iphone5
{
    public :
        void Show_S()
        {
            cout<<"Iphone5S { Color : "<<Color<<" \\ Prize : "<<Prize<<" }"<<endl;
        }
};

void output(int cas_id,Iphone5 five,Iphone5S fives)
{
    cout<<"Case #"<<cas_id<<":"<<endl;
    five.Show();
    fives.Show_S();
}
int main()
{
    int T;
    cin>>T;
    int cas=1;
    while(T--)
```

```
    {
        Iphone5 iphone5;
        Iphone5S iphone5s;
        char c[11];
        int p;
        cin>>c;
        cin>>p;
        iphone5.Set_Color(c);
        iphone5.Set_Prize(p);
        cin>>c;
        cin>>p;
        iphone5s.Set_Color(c);
        iphone5s.Set_Prize(p);
        output(cas++,iphone5,iphone5s);
    }
    return 0;
}
```

10.4 题号：11612

类的多重继承。

```
#include<iostream>
#include<cstdio>
using namespace std;
class Sony
{
    public :
        __int64 PrizeofSony;
};

class Iphone
{
    public :
        __int64 PrizeofIphone;
};

class Mi
{
```

```cpp
    public :
        __int64 PrizeofMi;
};

class Chuizi: public Iphone,public Sony, public Mi
{
    public :
        __int64 PrizeofChuizi;
        void Set_Prize(__int64 p1,__int64 p2,__int64 p3,__int64 p4)
        {
            PrizeofSony=p1;
            PrizeofMi=p2;
            PrizeofIphone=p3;
            PrizeofChuizi=p4;
        }
        bool cmp()
        {
            return PrizeofChuizi*100<(PrizeofSony+
                PrizeofMi+PrizeofIphone)*87;
        }
};

int main()
{
    int T;
    cin>>T;
    int cas=1;
    while(T--)
    {
        __int64 p1,p2,p3,p4;
        Chuizi chuizi;
        cin>>p1>>p2>>p3>>p4;
        chuizi.Set_Prize(p1,p2,p3,p4);
        cout<<"Case #"<<cas++<<": ";
        if(chuizi.cmp())
            cout<<"Yes"<<endl;
        else
```

```
            cout<<"No"<<endl;
    }
    return 0;
}
```

第 11 章 多态性

11.1 题号：11613

求平均分而已。

```
#include<iostream>
using namespace std;

class cs
{
    public:
        int sum;
        int avg_;
        void input_sum(int x)  { sum = x; }
        void input_avg(int y)  { avg_ = y; }
        int num_of_st()  { return sum/avg_; }
        int out_sum() { return sum; };
        int out_avg() { return avg_; };
};

class cs2 :public cs
{
    int num;
    public:
        void put_num(int x) { num = x; }
        int avg() { return sum/num; }
};
int main()
{
    cs *p;
```

```
        cs class1;
        cs grade;
        cs2 *dp;
        cs2 class2;
        int input;
        int sum_of_st = 0;
        int sum_of_scr_cs2=0;

        p = &class1;
        cin >> input;
        p -> input_sum( input );
        cin >> input;
        p -> input_avg( input );
        sum_of_st -= p -> num_of_st();
    sum_of_scr_cs2 = - p -> out_sum();

        p = &grade;
        cin >> input;
        p -> input_sum( input );
        cin >> input;
        p -> input_avg( input );
        sum_of_st += p -> num_of_st();
        sum_of_scr_cs2 += p -> out_sum();

        p = &class2;
        dp = &class2;
        p -> input_sum( sum_of_scr_cs2 );
        dp -> put_num( sum_of_st );
        cout<< dp -> avg() << endl;
        return 0;
    }
```

11.2 题号：11614

点到原点的距离。

```
#include<iostream>
#include<cmath>
#include<iomanip>
```

```cpp
using namespace std;
class point
{
    public:
            int  a;
            int  b;
            void input_a( int x ) { a = x; }
            void input_b( int y ) { b = y; }
            virtual double dis() { return sqrt (a * a + b * b); }
};
class point2:public point
{
    int c;
    public:
            void input_c( int z ) { c = z; }
            double dis() { return sqrt (a * a + b * b + c * c); }
};

int main()
{
   int input_x,input_y;
   int t;
   cin >> t ;
   char mark;
   while ( t -- )
   {
        cin >> mark;
        if( mark == 'b' )
        {
            point pt;
            point *p;
            p = &pt;
            cin >> input_x >> input_y;
            p -> input_a(input_x);
            p -> input_b(input_y);
            cout << fixed << setprecision(2) << p -> dis() << endl;
        }
        else
```

```cpp
        {
            int input_z;
            point2 pt;
            point2 *p;
            p = &pt;
            cin >> input_x >> input_y >> input_z;
            p -> input_a(input_x);
            p -> input_b(input_y);
            p -> input_c(input_z);
            cout << fixed << setprecision(2) << p -> dis() << endl;
        }
    }
    return 0;
}
```

11.3 题号：11615

不要交源程序。

```cpp
#include<iostream>
using namespace std;
int main()
{
    int t;
    while ( cin>>t &&t )
    {
        if( t ==2 )
        cout << "Banana" << endl;
        else
        cout << "Apple" << endl;
    }
    return 0;
}
```

11.4 题号：11616

少年敢再求一个距离吗？

```cpp
#include<iostream>
#include<cmath>
#include<iomanip>
```

```cpp
#include<algorithm>
using namespace std;

class point
{
    protected:
        int x,y;
    public:
        void set_point(int a,int b) { x = a,y = b; }
        virtual void dis() {cout << "Undefinition." << endl; }
};
class A_pt:public point
{
    public:
        void dis() { cout << fixed << setprecision(2) << sqrt(x * x +
            y * y) << endl; }
};
class B_pt:public point
{
    public:
        void dis() { cout << abs(x) + abs(y) << endl; }
};
int main()
{
    int t;
    cin >> t;
    while ( t-- )
    {
        point *p;
        A_pt pt1;
        B_pt pt2;
        int a,b;
        cin >> a >> b;
        p = &pt1;
        p -> set_point(a,b);
        p -> dis();
        p = &pt2;
        p -> set_point(a,b);
```

```
            p -> dis();
    }
    return 0;
}
```

第12章 输入/输出流

12.1

```
#include<iostream>
#include<fstream>
using namespace std;

int main()
{
    char str[300];
    int count=1;
    ifstream fin("c:\\a.txt");
    fin.getline(str,sizeof(str));
    while(!fin.eof())
    {
        count++;
        fin.getline(str,sizeof(str));
    }
    cout<<"The amount of lines of file is:"<<count<<endl;
    return 0;
}
```

12.2

```
#include<fstream>
using namespace std;

int main()
{
    ifstream fin("ride.in");
    ofstream fout("ride.out");
    char ch;
```

```
    while(fin.get(ch))
    {
        fout<<ch;
    }
    fin.close();
    fout.close();
    return 0;
}
```

12.3

```
#include<iostream>
#include<fstream>
using namespace std;

struct student
{
    int sno;
    char sname[10];
    float score;
};

int main()
{
    student st;
    int i,n;
    ofstream outfile("sinfo.dat",ios::out|ios::binary);
        if(!outfile)
        {
         cout<<"文件打开错误! ";
         exit(1);
        }
    cout<<"输入学生个数: ";
    cin>>n;
    for(i=0;i<n;i++)
    {
        cout<<"输入第"<<i+1<<"个学生的学号、姓名、成绩: ";
        cin>>st.sno;
        cin>>st.sname;
```

```
            cin>>st.score;
            outfile<<st.sno<<" "<<st.sname<<" "<<st.score<<endl;
        }
        outfile.close();
        return 0;
}
```

12.4

```
#include<iostream>
#include<iomanip>
using namespace std;

class RMB
{
public:
    RMB(double v =0.0):yuan(v){ jf =(v-yuan)*100+0.5; }
    operator double(){ return yuan+jf/100.0; }
    void display(ostream& out)
    {
        int n=1;    //为了用两字节的￥符号,得先求整数部分位数…
        for(int x=yuan; x/10; x/=10)
            n++;
        if(n>5)
        {
            out<<"*******.**"; return;
        }
        double t=yuan+jf/100.0;
        out <<setw(7-n) <<"￥" <<setiosflags(ios::fixed)
            <<setprecision(2) <<setw(n+3) <<t;
    }
protected:
    unsigned int yuan;
    unsigned int jf;
};

ostream& operator <<(ostream& oo, RMB& d)
{
    d.display(oo);
```

```
    return oo;
}

int main()
{
    RMB rmb(25000.52);
    cout <<"Initially rmb = " <<rmb <<"\n";
    rmb =2.0*rmb;
    cout <<"then rmb = " <<rmb <<"\n";
    return 0;
}
```

第 13 章 模板和标准库

13.1 题号：10543

Who's in the Middle。

```
#include <iostream>
#include <algorithm>

using namespace std;

int Arr[10000];

void prt(int n){
    printf("%d ",n);
}

int main(){
    int n;
    scanf("%d",&n);
    for(int i=0;i<n;++i)scanf("%d",Arr+i);

    //STL 通用算法，部分排序，注意区间都是前闭后开
    partial_sort(Arr,Arr+n/2+1,Arr+n);
```

```
        printf("%d\n",Arr[n/2]);
        return 0;
}
```

13.2 题号：10595

Civil War。

```cpp
#include <iostream>
#include <set>
#include <string>
using namespace std;

struct _t{
    int v;
    string name;
    _t():v(0){}
    _t(int a,string const&s):v(a),name(s){}
};

bool operator < (_t const&l,_t const&r){
    return l.v > r.v || l.v == r.v && l.name > r.name;
}

int main(){
    int nofkase;
    scanf("%d",&nofkase);
    for(int kase=1;kase<=nofkase;++kase){
        printf("Case %d:\n",kase);

        int n;
        scanf("%d",&n);

        set<_t> s;

        while(n--){
            char cmd[21];
            scanf("%s",cmd);
            if ( 'N' == *cmd ){
```

```
            char name[21];
            int v;
            scanf("%s%d",name,&v);
            s.insert(_t(v,name));
        }else{
            set<_t>::iterator it = s.begin();
            ++it;

            if ( s.begin()->v == it->v ){
                printf("%s %s\n",s.begin()->name.c_str(),
                    it->name.c_str());
                s.erase(s.begin());
                s.erase(s.begin());
                continue;
            }

            _t t(s.begin()->v,s.begin()->name);
            t.v -= it->v;
            printf("%s\n",it->name.c_str());
            s.erase(s.begin());
            s.erase(s.begin());
            s.insert(t);
        }
    }
    return 0;
}
```

13.3 题号：10895

简单集合运算。

```
#include <algorithm>
#include <iostream>
#include <iterator>
#include <string>
#include <sstream>

using namespace std;
```

```cpp
string string2int(int n){
    stringstream ss;
    ss<<n;
    return ss.str();
}

char A[30],B[30];

int main(){
    string::iterator it0,it1,it2;
    ostream_iterator<char> out(cout);
    int kase = 0;
    while( 1 ){
        char ch = getchar();
        if ( EOF == ch ) return 0;
        ungetc(ch,stdin);

        gets(A);gets(B);
        string a(A),b(B);
        string bing(a+b);
        string jiao(bing);
        string cha(bing);

        it0 = set_union(a.begin(),a.end(),b.begin(),b.end(),
           bing.begin());
        it1 = set_intersection(a.begin(),a.end(),b.begin(),b.end(),
           jiao.begin());
        it2 = set_difference(a.begin(),a.end(),b.begin(),b.end(),
           cha.begin());
        ++kase;

        string ca("Case ");
        ca += string2int(kase);
        ca += ":\n";

        copy(ca.begin(),ca.end(),out);

        copy(bing.begin(),it0,out);
```

```
        *out = '\n';
        ++out;

        copy(jiao.begin(),it1,out);
        *out = '\n';
        ++out;

        copy(cha.begin(),it2,out);
        *out = '\n';
        ++out;
    }
    return 0;
}
```

13.4 题号：10898

有序序列的合并。

```
#include <iostream>
#include <algorithm>
using namespace std;

int Arr[2000002*2];

int main(){
    int m,n;
    while( EOF != scanf("%d",&m) ){
        int i;
        for(i=0;i<m;++i)scanf("%d",Arr+i);

        scanf("%d",&n);
        for(;i<m+n;++i)scanf("%d",Arr+i);

        inplace_merge(Arr,Arr+m,Arr+m+n);

        if ( 0 == m && 0 == n ) {
            printf("\n");
            continue;
        }
```

```
        printf("%d",*Arr);
        for(i=1;i<m+n;++i)printf(" %d",Arr[i]);
        printf("\n");
    }
    return 0;
}
```

13.5 题号：10919

多关键字排序。

```
#include <iostream>
#include <algorithm>
using namespace std;

struct node_t{
    int d;
    int h;
    node_t(int x=0,int y=0):d(x),h(y){}
}Node[1000000+1];

bool operator < (node_t const&l,node_t const&r){
    return l.d > r.d || ( l.d==r.d && l.h>r.h );
}

inline bool check(int index,int d,int h){
    return d>Node[index].d || (d==Node[index].d && h>=Node[index].h);
}


//使用递归二分查找，溢出了
//改写成非递归
int bin_search(int n,int d,int h){
    int left = 0,right = n;
    while( right - left > 1 ){
        int mid = (left+right) / 2;
        if ( check(mid,d,h) ) right = mid;
        else            left = mid;
    }
```

```
        if ( check(left,d,h) ) return left;
        return right;
}


int main(){
    int n;
    scanf("%d",&n);

    for(int i=0;i<n-1;++i)scanf("%d%d",&Node[i].d,&Node[i].h);

    sort(Node,Node+n-1);

    int t;
    int d,h;
    scanf("%d",&t);
    for(int i=0;i<t;++i){
        scanf("%d%d",&d,&h);
        printf("%d\n",1+bin_search(n-1,d,h));
    }

    return 0;
}
```

13.6 题号：11068

义贼的问题。

```
#include <iostream>
#include <algorithm>
#include <vector>
using namespace std;

int main(){
    int n;
    while( cin>>n && n ){
        vector<int> v(n);
        for(int i=0;i<n;++i) cin>>v[i];
        vector<int>::iterator eit = unique(v.begin(),v.end());
```

```
            if ( v.begin() == eit ) {
                cout<<endl;
                continue;
            }

            cout<<*v.begin();
            for(vector<int>::iterator it = v.begin()+1;it!=eit;++it)
                cout<<' '<<*it;
            cout<<endl;
        }
        return 0;
    }
```

13.7 题号：11208

Stock Prices。

```
#include <iostream>
#include <map>
#include <algorithm>
using namespace std;

int Lowest[101];
int Highest[101];

int main(){
    int n,k1,k2;
    int kase = 0;
    while( scanf("%d%d%d",&n,&k1,&k2) ){
        if ( 0 == n && 0 == k1 && 0 == k2 )return 0;

        //value作为键，天数作为值
        multimap<int,int> v2d;
        for(int i=1;i<=n;++i){
            int x;
            scanf("%d",&x);
            v2d.insert(make_pair(x,i));
        }
```

```
            multimap<int,int>::const_iterator it=v2d.begin();
            for(int i=0;i<k1;++i,++it)
                Lowest[i] = it->second;

            it=v2d.end();--it;
            for(int i=0;i<k2;++i,--it)
                Highest[i] = it->second;

            sort(Lowest,Lowest+k1);
            sort(Highest,Highest+k2);

            //输出
            ++kase;
            printf("Case %d\n%d",kase,Lowest[0]);
            for(int i=1;i<k1;++i)
                printf(" %d",Lowest[i]);
            printf("\n%d",Highest[k2-1]);
            for(int i=k2-2;i>=0;--i)
                printf(" %d",Highest[i]);
            printf("\n");
        }
        return 0;
    }
```

13.8 题号：11294

剔除重复元素。

```
#include <iostream>
#include <algorithm>
using namespace std;
int main(){
    int t;
    scanf("%d",&t);
    while(t--){
        int n;
        scanf("%d",&n);
        int a[10005];
        for(int i=0;i<n;++i)scanf("%d",a+i);
```

```
    sort(a,a+n);
    int *p = unique(a,a+n);
    printf("%d ",p-a);
    for(int* x=a;x!=p;++x)printf("%d ",*x);
    printf("\n");
  }
}
```

13.9　题号：11298

ACM 比赛排名。

```
#include <iostream>
#include <algorithm>
using namespace std;
struct _t{
    string name;
    int cnt;
    int pen;
};
bool operator < (_t const& l,_t const& r){
    if ( l.cnt != r.cnt ) return l.cnt > r.cnt;
    if ( l.pen != r.pen ) return l.pen < r.pen;
    return l.name < r.name;
}
_t A[505];
int main(){
    int n;
    cin>>n;
    for(int i=0;i<n;++i)
        cin>>A[i].name>>A[i].cnt>>A[i].pen;
    sort(A,A+n);
    for(int i=0;i<n;++i)
        cout<<A[i].name<<' '<<A[i].cnt<<' '<<A[i].pen<<endl;
    return 0;
}
```

13.10 题号：11617

翻译文章。

```
#include<iostream>
#include<stdio.h>
#include<string.h>
#include<string>
#include<map>
#include<iterator>

using namespace std;

int main()
{
    map<string,string>m;
    map<string,string>::iterator it;
    char c;
    char s[11],s1[11];
    int i;
    while(scanf("%s",s) && strcmp(s,"END") != 0)
    {
        if(strcmp(s,"START") == 0) continue;
        scanf("%s",s1);
          m.insert(pair<string,string>(string(s),string(s1)));
    }
    i = 0;
    getchar();
    while(c = getchar())
    {
        if(c>='a'&&c<='z')
            s[i++]=c;
        else if(c=='S')
        {
            scanf("%s",s);
            getchar();
            i=0;
        }
        else if(c=='E')
```

```
            {
                scanf("%s",s);
                getchar();
                break;
            }
            else
            {
                s[i]='\0';
                it=m.find(string(s));
                if(it!=m.end())
                    printf("%s",(it->second).data());
                else
                    printf("%s",s);
                i=0;
                printf("%c",c);
            }
        }
        return 0;
    }
```

第四部分　模拟试卷

笔试模拟试卷（1）

课程代码：　　　考核方式：闭卷　　　考试时间：120 分钟　　　试卷类型：A

题　号	一	二	三	四	五	六	总分	合分人	复查人
应得分	30	12	16	12	10	20	100		
实得分									

一、单选题（在本题的每个小题的备选答案中，只有一个答案是正确的，请把你认为正确的答案的题号填入题干的括号内。多选不给分。每题 2 分，共 30 分。）

得分	评卷人	复查人

1. C++源文件的扩展名为（　　　）。
  ① cpp　　　　　　　　　　　② c
  ③ txt　　　　　　　　　　　④ exe
2. 下面关于类的描述，（　　　）是错误的。
  ① 类是抽象数据类型的实现　　② 类是具有共同行为的若干对象的统一描述体
  ③ 类是创建对象的模板　　　　④ 类就是 C 语言中的结构类型
3. 运算符+、=、*、<=中，优先级最高的运算符是（　　　）。
  ① +　　　　　　　　　　　　② =
  ③ *　　　　　　　　　　　　④ <=
4. 下列说法正确的是（　　　）。
  ① cout<<" \n "是一条语句，它能在屏幕上显示" \n "　　② \68 代表的是字符 D
  ③ 1E+5 的写法正确，它表示整型常量　　　　　　　　 ④ 0x10 相当于 020
5. 在下列原型所示的 C++函数中，按"传值"方式传递参数的是（　　　）。
  ① void f2(int *x);　　　　　　② void f1(int x);
  ③ void f3(const int *x);　　　　④ void f4(int &x);
6. 下列语句错误的是（　　　）。
  ① const int a;　　　　　　　　② const int a=10;
  ③ const int *point=0;　　　　　④ const int *point=new int(10);
7. 下列字符串不能作为 C++标识符使用的是（　　　）。
  ① WHILE　　　　　　　　　　② user
  ③ _lvar1　　　　　　　　　　　④ 9stars
8. 有如下程序：
  `#include<iostream.h>`

```
void main(){
    int sum;
    for(int i=0;i<6;i+=3){
        sum=i;
        for(int j=i;j<6;j++)
            sum+=j;
    }
    cout<<sum<<endl;
}
```
运行的结果是（　　）。

① 3　　　　　　　　　　　　② 10

③ 12　　　　　　　　　　　　④ 15

9. 下列语句中，正确的是（　　）。

① char mystring="Hello-world!";　　② char *mystring="Hello-world!";

③ char mystring[11]="Hello-world!";　　④ char mystring[12]="Hello-world!";

10. 若已经声明了函数原型"void fun(int a,double b=0.0);"，则下列重载函数声明中正确的是（　　）。

① void fun(int a=90,double b=0.0);　　② int fun(int a,double B);

③ void fun(double a,int B);　　④ bool fun(int a,double b=0.0);

11. 对类的构造函数和析构函数描述正确的是（　　）。

① 构造函数不能重载，析构函数可以重载

② 构造函数可以重载，析构函数不能重载

③ 构造函数不能重载，析构函数也不能重载

④ 构造函数可以重载，析构函数也可以重载

12. 下面关于静态数据成员的描述中，正确的是（　　）。

① 静态数据成员是类的所有对象共享的数据

② 静态数据成员不能通过类的对象调用

③ 类的不同对象有不同的静态数据成员值

④ 类的每一个对象都有自己的静态数据成员

13. 下列运算符中，（　　）运算符在C++中不能用友元函数重载。

① &&　　　　　　　　　　　　② ==

③ =　　　　　　　　　　　　　④ +

14. 派生类的对象对它的基类成员中（　　）是可以访问的。

① 公有继承的保护成员　　　　② 公有继承的私有成员

③ 私有继承的公有成员　　　　④ 公有继承的公有成员

15. 编译时的多态性可以通过（　　）得到。

① 重载函数和析构函数　　　　② 运算符和函数重载虚函数

③ 虚函数　　　　　　　　　　④ 虚函数和引用

二、填空题（每空 1 分，共 12 分）

1. 若有定义：int b=7; float a=2.5,c=4.7;，则下面表达式的值为_____。

a+(int)(b/3*(int)(a+c)/2)%4

2. 执行下面程序段后，k 值是_____。

k=1; n=263;
   do{ k*=n%10; n/=10;}while(n);

3. 已知字母 a 的 ASCII 码为十进制数 97，ch 为字符变量，则 ch='a'+'8'−'3'的值为_____。

4. 若有定义 static int b[4][3]={{1,2},{1},{4,6,8},{5,7}}，则初始化后 b[1][2]和 b[2][1]的值分别为_____、_____。

5. 将字符串 s2 连接到字符串 s1 后的函数调用是_____。

6. 设 int a[5]={10,20,30,40,50};int *p=&a[0];表达式++*p+*(a+3)的值是_____。

7. "判断整数 a 是否在闭区间[−10,10]中"的 C++语言表达式是_____。

8. 生成一个派生类对象时，先调用_____的构造函数，然后调用_____的构造函数。

9. 在一个抽象类中，一定包含有_____。

10. 有如下递归函数：

```
int Fun(int n){
    if(n<=1) return 1;
    _____
}
```

请补充完整，使得函数 Fun 能够正确计算形参 n 的阶乘。

三、写出下列程序的运行结果（每题 4 分，共 16 分）

1. 程序如下：

```
#include<iostream>
using namespace std;
struct stu
{
    int x;
    int *y;
}*p;
int dt[4]={10,20,30,40};
struct stu a[4]={50,&dt[0],60,&dt[1],70,&dt[2],80,&dt[3]};
int main()
```

```
    {
        p=a;
        cout<<++p->x<<"\n";
cout<<(++p)->x<<",";
        cout<<++(*p->y)<<endl;
        return 0;
    }
```
运行结果：

2. 程序如下：
```
    #include<iostream>
    using namespace std;
    int main()
    {
        int n=0,j=0,s=0;
        for(int i=8;i<=20;i++)
        {
            n++;
            i++;
            if(i%4==0)
                j++;
            else
                s++;
        }
        cout<<"n="<<n<<'\n'<<"j="<<j<<'\n'<<"s="<<s;
        return 0;
    }
```
运行结果：

3. 程序如下：
```
    #include<iostream.h>
    #include<string.h>
    class mystr
    {
        char string[81];
        int len;
    public:
        mystr(char *s){strcpy(string,s);}
        void getvalue(char *s,int &n){strcpy(s,string);n=len;}
```

```cpp
        friend mystr operator+(mystr a,mystr b)
        {
            mystr c(" ");
            strcpy(c.string,a.string);
            strcat(c.string,b.string);
            c.len=strlen(c.string);
            return c;
        }
};
int main()
{
    char str[81];
    int L;
    mystr a("Hello "),b("everybody!"),c(" ");
    c=a+b;
    c.getvalue(str,L);
    cout<<str<<endl<<"Len="<<L<<endl;
    return 0;
}
```

运行结果：

4. 程序如下：

```cpp
#include<iostream>
#include<iomanip>
using namespace std;
#define N 3
int main()
{
    int a[N][N];
    int i,j,k,n=1;
    for(k=0;k<(N+1)/2;k++)
    {
        for(i=k,j=k;j<N-1-k;j++)
        {
            a[i][j]=n;n++;
        }
        for(;i<N-1-k;i++)
        {
```

```
                a[i][j]=n;n++;
            }
            for(;j>k;j--)
            {
                a[i][j]=n;
                n++;
            }
            for(;i>k;i--)
            {
                a[i][j]=n;n++;
            }
        }
        if(N%2!=0)
            a[i][j]=n;
        for(i=0;i<N;i++)
        {
            cout<<"\n";
            for(j=0;j<N;j++)
                cout<<a[i][j]<<setw(4);
        }
        cout<<"\n";
        return 0;
    }
```

运行结果：

四、程序改错（每题只有两处错误，在错误处画线，并写出正确语句。每题 4 分，共 12 分）

得分	评卷人	复查人

1. 下面程序为实现两个数的互换。
```
#include<iostream>
using namespace std;
void swap(int x,int y)
{
int temp;
    temp=x;
    x=y;
    y=temp;
}
int main()
```

```cpp
{
    int x=5,y=6;
    cout<<x<<"  "<<y<<endl;
    swap(&x,&y);
    cout<<x<<"  "<<y<<endl;
    return 0;
}
```

2. 下面程序输出结果为："OECIA"。

```cpp
#include<iostream>
using namespace std;
#define M 14
#define N 7
int main()
{
    char a[M]="I Love China",b[N];
    int i,j=0;
    for(i=1;i<=M;i=i+2)
    {
        if(a[i]>='a' && a[i]<='z')
        {
            b[j]=a[i]-32;
            j++;
        }
        else
        {
            b[j]=a[i];
            j++;
        }
    }
    cout<<"\n 数组 b 为: "<<b[N];
    return 0;
}
```

3. 下列程序实现类的定义和使用。

```cpp
#include<iostream>
using namespace std;
class Point
{
private:
    int x,y;
    void init(int a,int b){x=a;y=b;}
```

```
        void show(){cout<<"x=  "<<x<<"    y="<<y<<endl;}
};
int main()
{
    Point a(24,50);
    a.show();
    return 0;
}
```

五、程序填空（每空 2 分，共 10 分）

得分	评卷人	复查人

1. 以下是建立链表函数。

```
struct student *creat(void)
{
    struct student  *head;
    struct student *p1,*p2;
    n=0;
       ①                       
    scanf("%d,%f",&p1->num,&p1->score);
    head=NULL;
    while(p1->num!=0)
    {
        n=n+1;
        if(n==1)
            head=p1;
        else
           ②                              
        p2=p1;
        p1=(struct student *)malloc(LEN);
        scanf("%d,%f",&p1->num,&p1->score);
    }
    p2->next=NULL;
       ③                         
}
```

2. 下面程序的运行结果是 10。

```
#include<iostream>
using namespace std;
class Amount
{
```

```
        int amount;
    public:
        Amount(int n=0):amount(n){}
        int getAmount(){return amount;}
        Amount &operator+=(Amount a);
};
Amount &Amount::operator+=(Amount a)
{
    amount+=a.amount;
    ①_____
}
int main()
{
    Amount x(3),y(7);
    x+=y;
    ②_____
    return 0;
}
```

六、编程题（每题 10 分，共 20 分）

得分	评卷人	复查人

1. 已知 10 个学生的成绩分别为 42、65、80、74、36、44、28、65、94、72，编写程序完成以下功能（三个功能都用函数实现）：

①对学生成绩采用插入排序法对其按从高到低进行排序，并输出结果。
②求出学生的不及格人数。
③求出学生的平均分。

2. 声明一个抽象类 shape（虚形），并由它派生出两个类 Rectangle（实形正方形）与 Circle（实形园），它们都有求面积和求周长的函数。

笔试模拟试卷（2）

课程代码：　　　考核方式：闭卷　　　考试时量：120分钟　　　试卷类型：B

题　号	一	二	三	四	五	六	总分	合分人	复查人
应得分	30	12	16	12	10	20	100		
实得分									

一、单选题（在本题的每个小题的备选答案中，只有一个答案是正确的，请把你认为正确的答案的题号填入题干的括号内。多选不给分。每题2分，共30分。）

得分	评卷人	复查人

1. C/C++语言规定：在一个源程序中，main 函数的位置（　　）。
  ① 必须在最开始　　　　　　　　② 必须在系统调用的库函数的后面
  ③ 可以任意　　　　　　　　　　④ 必须在最后

2. 若有以下定义，则能使值为3的表达式是（　　）。
  int k=7,x=12;
  ① x%=(k%=5)　　　　　　　　　② x%=(k-k%5)
  ③ x%=k-k%5　　　　　　　　　 ④ (x%=k)-(k%=5)

3. 当 a=1，b=3，c=5，d=4 时，执行完下面一段程序后 x 的值是（　　）。
```
       if(a<b)
         if(c<d) x=1;
         else if(a<c)
                if(b<d) x=2;
                else x=3;
              else x=6;
       else x=7;
```
  ① 1　　　　　　　　　　　　　　② 2
  ③ 3　　　　　　　　　　　　　　④ 6

4. 以下函数声明形式正确的是（　　）。
  ① double fun(int x, int y)　　　　② double fun(int x; int y);
  ③ double fun(int x,y)　　　　　　④ double fun(int x,y);

5. 若用数组名作为函数调用的实参，则传递给形参的是（　　）。
  ① 数组第一个元素的值　　　　　　② 数组的首地址
  ③ 数组中全部元素的值　　　　　　④ 数组元素的个数

6. 以下对二维数组 a 的正确说明是（　　　）。
   ① int a[3][];　　　　　　　　　② float a(3,4);
   ③ double a[1][4];　　　　　　　④ int a(3)(4);
7. 若已定义 int a=5; 下面对（1）、（2）两条语句的正确解释是（　　　）。
   （1）int *p=&a;　　　　（2）*p=a;
   ① 语句（1）和语句（2）中的*p 含义相同，都表示给指针变量 p 赋值
   ② 语句（1）和语句（2）语句的执行结果都是把变量 a 的地址值赋给指针变量 p
   ③ 语句（1）在对 p 进行说明的同时进行初始化，使 p 指向 a
      语句（2）将变量 a 的值赋给指针变量 p
   ④ 语句（1）在对 p 进行说明的同时进行初始化，使 p 指向 a
      语句（2）将变量 a 的值赋予*p
8. 在 C++中，编译系统自动为一个类生成默认构造函数的条件是（　　　）。
   ① 该类没有定义任何有参构造函数　　② 该类没有定义任何无参构造函数
   ③ 该类没有定义任何构造函数　　　　④ 该类没有定义任何成员函数
9. 下列关于运算符重载的叙述中，错误的是（　　　）。
   ① 有的运算符可以作为非成员函数重载
   ② 所有的运算符都可以通过重载而赋予新的含义
   ③ 有的运算符只能作为成员函数重载
   ④ 不得为重载的运算符函数的参数设置默认值
10. C++中，与实现运行时的多态性无关的是（　　　）。
    ① 重载函数　　　　　　　　　　　② 虚函数
    ③ 指针　　　　　　　　　　　　　④ 引用
11. 下列描述中，（　　　）是抽象类的特征。
    ① 可以说明虚函数　　　　　　　　② 可以进行构造函数重载
    ③ 可以定义友元函数　　　　　　　④ 不能说明其对象
12. 下面对静态数据成员的描述中，正确的是（　　　）。
    ① 静态数据成员是类的所有对象共享的数据
    ② 静态数据成员不能通过类的对象调用
    ③ 类的不同对象有不同的静态数据成员值
    ④ 类的每一个对象都有自己的静态数据成员
13. 下列运算符中，（　　　）运算符在 C++中不能用友元函数重载。
    ① &&　　　　　　　　　　　　　② ==
    ③ []　　　　　　　　　　　　　　④ +
14. 多继承派生类构造函数构造对象时，（　　　）被最先调用。
    ① 派生类自己的构造函数　　　　　② 虚基类的构造函数
    ③ 非虚基类的构造函数　　　　　　④ 派生类中子对象类的构造函数
15. 已知函数 f 的原型是 void f(int *a,long &b); 变量 v1、v2 的定义是：
    int v1;long v2; 下列调用语句正确的是（　　　）。

① f(v1,&v2);　　　　　　　　② f(v1,v2);
③ f(&v1,&v2);　　　　　　　④ f(&v1,v2);

二、填空题（每空 1 分，共 12 分）

1. C++中的标志符只能由字母、数字和下划线三种字符组成，且第一个字符_____。
2. 若 a 是 int 型变量，且 a 的初值为 6，则计算表达式后 a 的值为_____。
   a+=a-=a*a
3. 已知 x=43,ch='A',y=0；则表达式(x>=y&&ch<'B'&&!y)的值是_____。
4. C++将类继承分为_____、_____两种。
5. 如果定义 int a=2,b=3;float x=5.5,y=3.5; 则表达式(float)(a+b)/2+(int)x%(int)y 的值为_____。
6. 设 int a[5]={10,20,30,40,50};int *p=&a[0];表达式++*p+*(a+3)的值是_____。
7. 函数的递归分为_____和_____。
8. 若有以下定义，则变量 p 所占内存空间的字节数是_____。
   float *p;
9. 利用"对象名.成员变量"形式访问的对象成员仅限于被声明为_____的成员。
10. 执行以下程序段后，m 的值为_____。
    int a[2][3]={{1,2,3},{4,5,6}};
    int m,*p=&a[0][0];
    m=(*p)*(*(p+2))*(*(p+4));

三、写出下列程序的运行结果（每题 4 分，共 16 分）

1.
```
#include<iostream>
using namespace std;
struct stu
{
    int x;
    int *y;
}*p;
    int dt[4]={10,20,30,40};
    struct stu a[4]={50,&dt[0],60,&dt[1],70,&dt[2],80,&dt[3]};
int main()
{
```

```
        p=a;
        cout<<++p->x<<"\n";
        cout<<(++p)->x<<",";
        cout<<++(*p->y)<<endl;
        return 0;
    }
```
运行结果：

2.
```
    #include<iostream>
    #include<iomanip>
    using namespace std;
    int main()
    {
        int i,j,n;
        n=3;
        for(i=1;i<=n;i++)
        {
            cout<<setw(n-i+1);
            for(j=1;j<=2*i-1;j++)
                cout<<'#';
            cout<<endl;
        }
        return 0;
    }
```
运行结果：

3.
```
    #include<iostream>
    using namespace std;
    class count
    {
        static int n;
    public:
        count(){n++;}
        void show(){cout<<n<<endl;}
        ~count(){cout<<n<<endl;n--;}
    };
    int count::n=0;
```

```
    int main()
    {
        count a;
        a.show();
        {
            count b[4];
            b[3].show();
        }
        a.show();
        return 0;
    }
```
运行结果：

4.
```
    #include<iostream>
    using namespace std;
    class base
    {
        int i,j;
    public:
        void set(int a , int b)  { i = a; j = b; }
        void show()  { cout<<i<<" "<<j<<"\n";}
    };
    class derived : public base
    {
        int k;
    public:
        derived(int x)  { k=x; }
        void showk()  { cout<<k<<"\n"; }
    };
    int main()
    {
        derived ob(3);
        ob.set(1,2);
        ob.show();
        ob.showk();
        return 0;
    }
```

运行结果：

四、程序改错（每题只有两处错误，在错误处画线，并写出正确语句。每题 4 分，共 12 分）

得分	评卷人	复查人

1.
```
#include<iostream>
using namespace std;
class base
{
protected:
    int i,j;
public:
    void set(int a,int b){i=a;j=b;};
    void show(){cout<<i<<" "<<j<<"\n";};
};
class derived1:private base
{
    int k;
public:
    void setk(){k=i*j;};
    void showk(){cout<<k<<"\n";};
};
class derived2: public derived1
{
int m;
public:
    void setm(){m=i-j;};
    void showm(){cout<<m<<"\n";};
};
int main()
{
    derived1 ob1;
    derived2 ob2;
    ob1.set(1,2);
ob1.show();
    ob2.set(3,4);
ob2.show();
return 0;
```

}

2.
```
#include<iostream>
using namespace std;
int main()
{
    int i,j,a[3][2]={1,2,3,4,5,6};
    int *p=a;
    for(i=0;i<3;i++)
    {
        for(j=0;j<2;j++)
            cout<<" "<<*(*(p+i)+j);  //采用指针方式访问p所指向的二维数组
        cout<<endl;
    }
    return 0;
}
```

3.
```
#include<iostream>
using namespace std;
class one
{
    int a1,a2;
public:
    void initial(int x1,int x2)
    {
        a1=x1;
        a2=x2;
    }
};
int main()
{
    one data(2,3);
    cout<<data.a1<<endl;
    cout<<data.a2<<endl;
    return 0;
}
```

五、程序填空（每空2分，共10分）

得分	评卷人	复查人

1.
```
#include<iostream>
using namespace std;
#define N 3
int main()
{
    char *str[N];
    int   i;
    for(i=0;i<N;i++)
    {
        ①_____
        cin>>str[i];
    }
    for(i=0;i<N;i++)
        ②_____    //要求输出后换行
    return 0;
}
```

2. 下面程序用"两路合并法"把两个已按升序排列的数组合并成一个升序数组。
```
#include<iostream>
using namespace std;
int main()
{
    int a[3]={5,9,19};
    int b[5]={12,24,26,37,48};
    int c[10],i=0,j=0,k=0;
    while(i<3&&j<5)
        if(①_____)
        {
            c[k]=b[j];
            k++;
            j++;
        }
        else
        {
            c[k]=a[i];
            k++;
```

```
            i++;
        }
    while(②_____)
    {
        c[k]=a[i];
        i++;
        k++;
    }
    while(③_____)
    {
        c[k]=b[j];
        k++;
        j++;
    }
    for(i=0;i<k;i++)
        cout<<c[i]<<"  ";
    return 0;
}
```

六、编程题（每题10分，共20分）

1. 编写程序，将输入的一行字符进行加密和解密。加密时，每个字符依次反复加上"4962873"中的数字，如果范围超过 ASCII 码的 032（空格）~122（"z"），则进行模运算。解密与加密的顺序相反。编制加密与解密函数，打印各个过程的结果。

2. 设计一个圆类 circle 和一个桌子类 table，另设计一个圆桌类 roundtable，它是从前两个类派生的，要求输出一个圆桌的高度 height、面积和颜色等数据。圆类 circle 的数据成员是半径 radius，桌子类 table 的数据成员是高度 height。输入数据是：高度为 0.8，半径为 1.2，颜色为"黑色"。

笔试模拟试卷（3）

课程代码：　　　　考核方式：闭卷　　　考试时间：120 分钟　　　试卷类型：A

题号	一	二	三	四	总分	合分人	复查人
应得分	40	24	18	18	100		
实得分							

一、单选题（每小题 2 分，共 40 分，请在下方的答题处作答。）

得分	评卷人	复查人

1. 计算机能够直接识别和接受的二进制代码集合被称为（　　）。
    ① 机器语言　　　② 符号语言　　　③ C 语言　　　④ C++语言
2. 下列选项中，属于 C++合法的标志符的是（　　）。
    ① A_2　　　　　② 2A　　　　　③ 2_A　　　　　④ 2
3. 下列选项中，不属于 C++关键字的是（　　）。
    ① int　　　　　② float　　　　③ double　　　　④ real
4. 下列选项中，数值最大的是（　　）。
    ① 1000　　　　② 1E3　　　　　③ 0x1000　　　　④ 01000
5. 下列有关数组的定义和初始化，不正确的是（　　）。
    ① int a[] = {1000,2000};　　　　② int a[3] = {1000,2000};
    ③ int a[3] = 1000;　　　　　　　④ int a[3] = {1000};
6. 下列程序片段的输出结果是（　　）。
    char s[] = "china"; cout<<s+3<<endl;
    ① china　　　　② hina　　　　③ ina　　　　　④ na
7. 下列程序片段的输出结果是（　　）。
    ```
    int a=1, b;
    for(b=1;b<=10;b++){
        if (a>=8)break;
        if(a%2==1){a+=5;continue;}
        a-=3;
    }
    cout<<b<<endl;
    ```
    ① 6　　　　　　② 5　　　　　　③ 4　　　　　　④ 3
8. 下列程序片段的输出结果是（　　）。

```
int a[]={2,4,6,8,10},y=0,x,*p;p=&a[1];
for(x=1;x<3;x++) y+=p[x];cout<<y<<endl;
```
① 16　　　　　② 15　　　　　③ 14　　　　　④ 13

9. 下列程序片段的输出结果是（　　）。
```
int x=1,y=0;
if(!x) y++;
else if(x==0)
if(x) y+=2;
else y+=3;
cout<<y<<endl;
```
① 3　　　　　② 2　　　　　③ 1　　　　　④ 0

10. 一个递归函数定义如下，当调用 fun(3)时，该函数被调用的总次数是（　　）。
```
int fun(int k){
    if(k<1) return 0;
    else if(k==1) return 1;
    else return fun(k-1)+1;
}
```
① 3　　　　　② 2　　　　　③ 1　　　　　④ 0

11. 下列关于函数重载的选项，说法正确的是（　　）。
① 函数参数的数量不能作为重载的依据
② 函数参数的类型不能作为重载的依据
③ 函数的类型不能作为重载的依据
④ 函数参数的名字可以作为重载的依据

12. 下列 C++的运算符中，优先级最低的是（　　）。
① ==（等于号）　　② =（赋值符号）　　③ +（加号）　　④ &&（逻辑与）

13. 下列 C++的运算符中，属于左结合的是（　　）。
① !（逻辑非）　　② ++（前缀自增）　　③ +（正号）　　④ >（大于号）

14. 下列程序片段的输出结果是（　　）。
```
int b[3][3]={0,1,2,0,1,2,0,1,2},i,j,t=1;
for(i=0;i<3;i++)
for(j=i;j<=i;j++)
t+=b[i][b[j][i]];
cout<<t<<endl;
```
① 1　　　　　② 3　　　　　③ 4　　　　　④ 9

15. 下列程序的运行结果是（　　）。
```
#define SUB(a) (a)-(a)
int main(){ int a=2,b=3,c=5,d; d=SUB(a+b)*c; cout<<d<<endl; return 0;}
```
① −20　　　　② −12　　　　③ 10　　　　④ 0

16. 下列关于循环的选项，说法正确的是（　　）。
① for 循环的功能最强大　　　　　　② do-while 循环的功能最弱
③ 使用 if 加 break 可以实现循环条件控制　　④ while 循环的条件表达式可以省略

17. 表达式 a+=a-=a=9 的计算结果是（　　）。
① 18　　　　② 9　　　　③ -9　　　　④ 0

18. 下列程序片段的输出结果是（　　）。
```
int a=1,b=2;
while(a<6) {b+=a; a+=2; b%=10;}
cout<<a<<","<<b<<endl;
```
① 5,11　　　② 7,1　　　③ 7,11　　　④ 6,1

19. 下列程序片段的输出结果是（　　）。
```
int y=10;
while(y--);
cout<<y<<endl;
```
① 10　　　　② 1　　　　③ 0　　　　④ -1

20. 下列程序的运行结果是（　　）。
```
int f(int m){
    static int n=0;n+=m;return n;
}
int main(){
    int n=0;
    cout<<f(++n)<<endl;
    cout<<f(n++)<<endl;
    return 0;
}
```
① 12　　　　② 11　　　　③ 23　　　　④ 33

二、程序修改题（每小题 6 分，共 24 分。打星号的行有错误，请就近写出正确代码，每道题有且只有两处错误，每处错误仅涉及一行，其他行不得进行改动。）

得分	评卷人	复查人

1. 函数 fun 的功能是：从整数 10 到 55 之间，选出能被 3 整除且有一位是 5 的那些数，答案保存在 b 数组中，这些数的数量由 return 返回。其中，特别要求局部变量 a2 保存十位上的数字，a1 保存各位上的数字。

```
int fun (int b[]){
    int k,a1,a2,i=0;
    for (k=10;k<=55;k++){
/************found************/a2=k%10;
```

```
            a1=k-a2*10;
            if((k%3==0 && a2==5)|| (k%3==0 && a1==5)){   b[i]=k;i++;}
        }
/*************found************/return k;
}
```

2. 函数 fun 的功能是计算调和级数。即计算 1+1/2+1/3+1/4+⋯+1/m。其中 m 是形参。

```
double fun( int m ){
    double t = 1.0;
    int i;
    for( i = 2; i <= m; i++ )
/**********found**********/t += 1/k;
/**********found**********/return k;
}
```

3. 函数 fun 的功能是：判断形参 n 是否是完数，是则返回 true，否则返回 false，而且，数组 a 保存 n 的因子，k 保存因子的个数。

```
bool fun(int n, int a[], int &k){
    int m=0, i, t = n;
/**********found**********/for( i=0; i<n; i++ )
        if(n%i==0) {  a[m]=i;  m++;  t=t - i;  }
    k=m;
/**********found**********/if ( t=0 )  return true;
    else  return false;
}
```

4. 函数 fun 的功能是计算 s=aa⋯aa-⋯-aaa-aa-a。其中 aa⋯aa 表示由 n 个 a 构成的正整数。a 和 n 是函数形参。例如，当 a、n 取 3、6 时，有
   s = 333333 - 33333 - 3333 - 333 - 33 - 3
其结果应为 296298。

```
int fun (int a, int n)
{ int j ;
/***************found***************/int s = 0, t = 1 ;
    for ( j = 0 ; j < n ; j++) t = t * 10 + a ;
    s = t ;
    for ( j = 1 ; j < n ; j++) {
/***************found***************/t = t % 10 ;
       s = s - t ;
    }
    return(s) ;
```

}

三、程序填空题（每小题9分，共18分。在画线处填入符合语法的内容以完成函数功能。）

1. 函数 fun 的功能是：将形参 n 各位上的偶数剔除，剩下的数按原顺序组成一个新的数，并返回。例如：输入 27638496，则应该返回 739。

```
int fun(int n){
    int x=0, i=1, t;
    while(n){
        t=*n % _____;
        if(t%2!= _____){ x=x+t*i;  i=i*10;}
        n /= 10;
    }
    return _____;
}
```

2. 函数 fun 的功能是：求 n 个数的平均值。

```
double fun(double x[],int n){
    double s = _____, ave;
    for(int i=0;i<n;++i) s += x[i];
    ave = s / _____;
    return _____;
}
```

四、程序设计题（每小题9分，共18分）

1. 编写程序实现以下功能：从键盘输入 3 个 int，输出其最小值。无需书写头文件和 using 声明，但必须书写主函数。

2. 编写程序实现以下功能：从键盘输入一个不带空格的字符串，长度最多为 80，输出其中英文小写字母的数量。无需书写头文件和 using 声明，但必须书写主函数。

笔试模拟试卷（4）

课程代码：　　考核方式：闭卷　　考试时间：120 分钟　　试卷类型：B

题号	一	二	三	四	总分	合分人	复查人
应得分	40	24	18	18	100		
实得分							

一、单选题（每小题 2 分，共 40 分，请在下方的答题处作答。）

得分	评卷人	复查人

1. 下列选项中，不属于高级程序设计语言的是（　　）。
   ① 机器语言　　② Java 语言　　③ C 语言　　④ C++语言
2. 下列选项中，属于C++合法的整型字面值的是（　　）。
   ① 02　　② 2H　　③ H2　　④ 1E3
3. 下列选项中，属于C++关键字的是（　　）。
   ① Integer　　② integer　　③ Int　　④ int
4. 下列选项中，数值最大的是（　　）。
   ① 1000L　　② 1E+3　　③ 1000.1　　④ 1000U
5. 下列有数组的定义和初始化，不正确的是（　　）。
   ① int a[] = {1000,2000};　　② int a[3] = {1000,2000};
   ③ int n=3, a[n];　　④ int a[3] = {1000};
6. 下列程序片段的输出结果是（　　）。
   char s[]="\\141\141abc\t"; cout<<strlen(s)<<endl;
   ① 14　　② 13　　③ 12　　④ 9
7. 下列程序片段的输出结果是（　　）。
   ```
   int i=0,a=0;
   while(i<20)
   { for( ; ; )
       if((i%10)==0) break;
       else i--;
     i+=11;
     a+=i;
   }
   cout<<a<<endl;
   ```

① 11　　　　② 21　　　　③ 32　　　　④ 33

8. 下列程序片段的输出结果是（　　）。

    int a=4, b=5, c=0, d; d=!a&&!b||!c;
    cout<<d<<endl;

① 3　　　　② 2　　　　③ 1　　　　④ 0

9. 下列程序片段的输出结果是（　　）。

    int b[3][3]={0,1,2,0,1,2,0,1,2},i,j,t=1;
    for(i=0;i<3;i++)
    for(j=i;j<=i;j++)
    t=t+b[i][b[j][j]];
    cout<<t<<endl;

① 9　　　　② 4　　　　③ 1　　　　④ 0

10. 下列程序片段中，while 循环的循环次数是（　　）。

    int i=0;
    while(i<10)
    { if(i<1) continue;
    if(i==5) break;
    i++;
    }

① 死循环　　　② 10　　　　③ 6　　　　④ 1

11. 下列关于函数的选项，说法正确的是（　　）。

    ① 函数不能进行递归调用
    ② 只有 main 函数才能调用 main 函数
    ③ 函数的声明可以不包括形参名
    ④ 函数形参的默认值必须从左往右

12. 下列 C++的运算符中，优先级最低的是（　　）。

    ① [ ]（方括号）　　② +=（复合赋值）　　③ +（加号）　　④ ||（逻辑或）

13. 下列 C++的运算符中，属于右结合的是（　　）。

    ① ==（等于号）　　② ++（后缀自增）　　③ +（正号）　　④ >（大于号）

14. 下列程序片段的输出结果是（　　）。

    int a=0,i;
    for(i=1;i<5;i++){
    switch(i)
    { case 0: case 3: a+=2;
    case 1: case 2: a+=3;
    default: a+=5;
    }
    }cout<<a<<endl;

① 31 ② 13 ③ 10 ④ 20

15. 若有定义：int a[8];则下列选项中不能表示 a[1]的地址的是（   ）。
① &a[0]+1 ② &a[1] ③ ++a[0] ④ a+1

16. 下列关于循环的选项，说法正确的是（   ）。
① for 循环的功能最强大
② do-while 循环的功能最弱
③ 使用 if 加 break 可以实现循环条件控制
④ while 循环的条件表达式可以省略

17. 表达式 a+=a-=a=9 的计算结果是（   ）。
① 18 ② 9 ③ -9 ④ 0

18. 下列程序片段的输出结果是（   ）。
```
int x=3,y=4,z=5;
cout<<(int)(!(x+y)+z-1&&y+z/2)<<endl;
```
① 0 ② 1 ③ 2 ④ 6

19. 下列程序片段的输出结果是（   ）。
```
int y=10;
while(y--);
cout<<y<<endl;
```
① 10 ② 1 ③ 0 ④ -1

20. 下列程序的运行结果是（   ）。
```
int a=3;
main(){
    int s=0;
    { int a=5;s+=a++;}
    s+=a++;
    cout<<s<<endl;
}
```
① 11 ② 10 ③ 8 ④ 7

二、程序修改题（每小题 6 分，共 24 分。打星号的行有错误，请就近写出正确代码，每道题有且只有两处错误，每处错误仅涉及一行，其他行不得进行改动。）

得分	评卷人	复查人

1. 函数 fun 的功能是：将字符串 s 中的字母转换为其字母表中的后续字母，即 A 变成 B，B 变成 C，但是 Z 要变成 A，z 要变成 a。
```
void fun (char *s)
{
/***********found***********/while(*s!='@')
```

```
        { if(*s>='A' && *s<='Z' || *s>='a' && *s<='z')
          { if(*s=='Z')  *s='A';
            else if(*s=='z')  *s='a';
            else              *s += 1;
          }
/**********found**********/(*s)++;
        }
    }
```

2. 函数 fun 的功能是：剔除字符串 s 中所有的非数字字母。
```
    void  fun(char *s)
    {  int  i,j;
       for(i=0,j=0; s[i]!='\0'; i++)
           if(s[i]<='0' || s[i]>='9')
/**********found**********/s[j]=s[i];
/**********found**********/s[j]="\0";
    }
```

3. 函数 fun 的功能是：按下列公式计算圆周率。

$$\frac{\pi}{2}=1+\frac{1}{3}+\frac{1}{3}\times\frac{2}{5}+\frac{1}{3}\times\frac{2}{5}\times\frac{3}{7}+\cdots$$

```
    double fun(double  eps){
        double  s,t;
        int   n=1;
        s=0.0;
/************found************/t=0;
        while( t>eps){
            s+=t;
            t=t * n/(2*n+1);
            n++;
        }
/************found************/return t;
    }
```

4. 函数 fun 的功能是：计算平方的倒数的和，即 y=1+1/(2*2)+1/(3*3)+…+1/(m*m)。其中，m 是形参，y 是返回值。
```
    double  fun ( int  m )
    {
        double  y = 1.0 ;
        int  i;
/***************found***************/for(i = 2 ; i < m ; i++)
```

```
/**************found**************/y += 1 / (i * i) ;
    return( y ) ;
}
```

三、程序填空题（每小题 9 分，共 18 分。在画线处填入符合语法的内容以完成函数功能。）

1. 函数 fun 的功能是：判断字符串形参 s 是否为回文。abcba 是回文，而 abcabc 不是回文。是就返回 true，否就返回 false。

```
bool fun(char *s)
{ char *lp,*rp;
  lp= _____ ;
  rp=s+strlen(s)-1;
  while((toupper(*lp)==toupper(*rp)) && (lp<rp) ) {lp++; rp _____ ; }
  if(lp<rp) return _____ ;
  else   return false;
}
```

2. 函数 fun 的功能是：求 n 个数的和。

```
double fun(double x[],int n){
double s = _____ ;
for(int i=0;i<n;++i)
    s += _____ ;
return _____ ;
}
```

四、程序设计题（每小题 9 分，共 18 分）

1. 编写程序实现以下功能：从键盘输入三个 int，输出其最大值。无需书写头文件和 using 声明，但必须书写主函数。

2. 编写程序实现以下功能：从键盘输入一个不带空格的字符串，长度最多为 80 个字符串，输出其中英文大写字母的数量。无需书写头文件和 using 声明，但必须书写主函数。

笔试模拟试卷（5）

课程代码：　　　考核方式：闭卷　　　考试时间：120 分钟　　　试卷类型：C

题号	一	二	三	四	总分	合分人	复查人
应得分	40	24	18	18	100		
实得分							

一、单选题（每小题 2 分，共 40 分，请在下方的答题处作答。）

得分	评卷人	复查人

1. 下列选项中，不属于高级程序设计语言的是（　　）。
   ① 机器语言　　② Java 语言　　③ C 语言　　④ C++语言

2. 下列选项中，属于 C++合法的字符串字面值的是（　　）。
   ① '\n'　　② '/n'　　③ "n"　　④ 'ABCD'

3. 下列选项中，属于 C++关键字的是（　　）。
   ① where　　② when　　③ while　　④ who

4. 下列选项中，数值最小的是（　　）。
   ① 1000L　　② 1E-3　　③ 0.01　　④ 01

5. 下列有字符串的定义和初始化，不正确的是（　　）。
   ① char a[] = "abcd";　　② char a[] = {"abcd"};
   ③ char a[4] = "abcd";　　④ char a[100] = "abcd";

6. 下列程序片段的输出结果是（　　）。
   ```
   for(int i=0;i<3;i++)
   switch(i)
   {
   case 1: cout<<i;
   case 2: cout<<i;
   default: cout<<i;
   };
   ```
   ① 011122　　② 012　　③ 012020　　④ 120

7. 下列程序片段的输出结果是（　　）。
   ```
   int a=5,b=4,c=3,d=2;
   if(a>b>c)cout<<d<<endl;
   else if((c-1>=d)==1)cout<<d+1<<endl;
   else cout<<d+2<<endl;
   ```

① 2　　　　　② 3　　　　　③ 4　　　　　④ 5

8. 下列程序片段的输出结果是（　　）。

    int i=0,s=0;
    do{if(i%2){i++;continue;}i++; s+=i;} while(i<7);
    cout<<s<<endl;

① 21　　　　② 12　　　　③ 16　　　　④ 28

9. 下列程序片段的输出结果是（　　）。

    int i=10,j=1; cout<<i--<<++j<<endl;

① 92　　　　② 102　　　③ 91　　　　④ 101

10. 下列程序的运行结果是（　　）。

    void f(int x,int y){int t;if(x<y){t=x; x=y; y=t;}}
    int main(){
    int a=4,b=3,c=5;
    f(a,b); f(a,c); f(b,c);
    cout<<a<<b<<c<<endl;
    return 0;
    }

① 345　　　② 543　　　③ 534　　　④ 435

11. 下列关于 return 的选项，说法不正确的是（　　）。

① return 后面可以接一个表达式
② 如果函数类型为 int，则 return 一个 double 值也是合法的
③ 不能 return 指针
④ 结构体类型的变量可以作为 return 的对象

12. 下列 C++的运算符中，优先级最高的是（　　）。

① *（乘号）　② +=（复合赋值）　③ +（加号）　④ !=（不等于）

13. 下列 C++的运算符中，属于右结合的是（　　）。

① =（赋值）　② ++（后缀自增）　③ /（除号）　④ ,（逗号）

14. 下列程序片段的输出结果是（　　）。

    char *s[]={"one","two","three"},*p = s[1];cout<<*(p+1)<<endl;

① w　　　　② wo　　　　③ one　　　　④ two

15. 下列程序片段的输出结果是（　　）。

    int m[][3]={1,4,7,2,5,8,3,6,9},i,j,k=2;
    for(i=0;i<3;i++) cout<<m[k][i];

① 456　　　② 258　　　③ 369　　　④ 789

16. 下列关于循环的选项，说法正确的是（　　）。

① for 循环的功能最强大　　　　　② do-while 循环的功能最弱
③ 使用 if 加 break 可以实现循环条件控制　　④ while 循环的条件表达式可以省略

17. 下列选项中，属于 C++合法整型字面值的是（　　）。

① 1011B　　　　② 0xFF　　　　③ ffH　　　　④ 0866

18. 下列程序片段的输出结果是（　　　）。

```
int x=3,y=4,z=5;
cout<<(int)(!(x+y)+z-1&&y+z/2)<<endl;
```

① 0　　　　② 1　　　　③ 2　　　　④ 6

19. 下列程序的运行结果是（　　　）。

```
int fun(int a, int b)
{ if(a>b) return(a);
  else return(b);
}
int main(){
int x=3,y=8,z=6,r;
r=fun(fun(x,y),2*z); cout<<r<<endl;return 0;
}
```

① 3　　　　② 12　　　　③ 8　　　　④ 6

20. 下列程序的运行结果是（　　　）。

```
#define F(x,y) (x)*(y)
int main(){ int a=3,b=4;
cout<<F(a++,b++)<<endl;return 0;
}
```

① 12　　　　② 15　　　　③ 16　　　　④ 20

得分	评卷人	复查人

二、程序修改题（每小题 6 分，共 24 分。打星号的行有错误，请就近写出正确代码，每道题有且只有两处错误，每处错误仅涉及一行，其他行不得改动。）

1. 函数 fun 的功能是：计算形参 num 各位上数字的积。

```
int fun(int num)
{
/************found************/ int k;
  do
  { k*=num%10;
/************found************/ num\=10;
  } while(num);
  return (k);
}
```

2. 函数 fun 的功能是：将形参字符串 a 逆序输出，但不改变 a 本身的内容。

```
/************found************/fun(char a)
```

```
{
if(*a)
    {
    fun(a+1)  ;
/************found************/cout<*a<endl;
    }
}
```

3. 函数 fun 的功能是：使用递归函数计算 Fibonacci 数列。所谓 Fibonacci 数列，是指：

$$F(n) = F(n-1) + F(n-2)$$

且 $F(1) = F(2) = 1$，所以 $F(7)=13$。

```
long fun(int g)
{
/************found************/ switch(g);
    {
case 0:return 0;
/************found************/case 1;
case 2:return 1;
    }
    return(fun(g-1)+fun(g-2));
}
```

4. 函数 fun 的功能是：将形参字符串 s 中的所有小写字母'c'删除。例如字符串为 aabbccAABBCC，则处理过后的字符串应为 aabbAABBCC。

```
void fun(char *s)
{
int i,j;
    for(i=j=0;s[i]!='\0';i++)
        if(s[i]!='c')
/************found************/s[j]=s[i];
/************found************/s[i]='\0';
}
```

三、程序填空题（每小题 9 分，共 18 分。在画线处填入符合语法的内容以完成函数功能。）

得分	评卷人	复查人

1. 函数 fun 的功能是：统计形参 n 的各位，出现数字 1、2、3 的次数。结果分别保存在全局变量 c1、c2、c3 中。

```
int  c1,c2,c3;
```

```
void fun(int n)
{   c1 = c2 = c3 = 0;
    while (n) {
        switch(_____)    {
          case 1:   c1++;_____;
          case _____:   c2++;break;
          case 3:   c3++;
        }
        n /= 10;
    }
}
```

2. 函数 fun 的功能是：求 n 个数的最大值。

```
double fun(double x[],int n){
double ret = _____;
for(int i=1;i<n;++i)
   if( ret _____ x[i] )
       ret = x[i];
return _____;
}
```

四、程序设计题（每小题 9 分，共 18 分）

得分	评卷人	复查人

1. 编写程序实现以下功能：从键盘输入三个 int，输出其中间的数值，即不大不小的值。无需书写头文件和 using 声明，但必须书写主函数。

2. 编写程序实现以下功能：从键盘输入一个不带空格的字符串，长度最多为 80 个字符串，输出其中数字字母的数量。无需书写头文件和 using 声明，但必须书写主函数。

上机实验考试模拟试卷（1）

1001　简单的题（题号 11519）

题目描述：虽然大家已经身经百战、参加过很多次考试了。但是在大学里参加期末考试应该还是第一次。所以这道题是一道非常简单的题。

输入两个 32 位整数，输出其和。

输入：输入有若干行，每行两个整数，用一个空格分隔。0 0 表示输入结束，且该行不需要再计算了。

输出：每行一个数，表示对应输入的和。

样例输入：

    1 2
    0 0

样例输出：

    3

1002　假期的长度（题号 11520）

题目描述：考完期末考试就该放假了，计算一下今年的假期有多长吧。

输入：输入多行，每行都包括四个数，即 y m d1 d2。第一个数表示年份，取值范围为[1900, 2050]。第二个数表示月份，取值范围为[1, 11]。第三个数表示假期第一天所在的日期。第四个数表示假期最后一天所在的日期。为了保证假期总是比较长，应保证假期的最后一天总在假期开始时的下一个月。所有日期输入肯定都是合法的。

输出：对每一行输入，给定一行输出。输出假期的长度，即假期一共有多少天。

样例输入：

    2014 1 30 1

样例输出：

    3

1003　第二大的数（题号 11521）

题目描述：古语有云："枪打出头鸟"，"出头的椽子先烂"，说明第一个站出来的人是最容易被发现的。不过自从有了计算机，第二个站出来的人也很容易被找出来。

输入 n 个整数，输出其中第二大的数。n 的范围为[2, 1000]，每个数的范围在[0, 1000]之间。特别说明，此处第二大指的是数值第二大。假设有三个数 1、4、4，则 1 就是第二大的那个数。输入保证不存在 n 个数都相等的情况。

输入：输入若干行，每行若干个整数，且之间用一个空格分隔。第一个数是 n，接下来还有 n 个数。

输出：每行一个数，表示输入的 n 个数中第二大的那个数。

样例输入：

```
    3 1 4 4
```
样例输出:
```
    1
```

1004 质数的和(题号 11522)

题目描述:给定区间[a, b],求该区间内所有质数的和。1≤a≤b≤30000。

输入:输入若干行,每行两个整数为 a 和 b。

输出:每行一个数为答案。

样例输入:
```
    1 10
```
样例输出:
```
    17
```

上机实验考试模拟试卷(2)

1001 简单的题 2015(题号 11618)

题目描述:每一届的师兄师姐都会碰到一道简单的题,2015 级也不例外。输入两个 32 位整数,输出其差。

输入:输入若干行,每行两个整数,用一个空格分隔。0 0 表示输入结束,且该行不需要再计算了。

输出:每行一个数,表示对应输入的差。

样例输入:
```
    2 1
    0 0
```
样例输出:
```
    1
```

1002 飞跃 2015(题号 11619)

题目描述:2015 年对 2015 级的同学来说意义重大,上半年是中学生,下半年就是大学生了。来计算一下中学生"距离"大学生还有多少天。给定 2015 年的两个日期,问相隔多少天?

输入:输入多行,每行都包括四个数,即 m1 d1 m2 d2。表示两个日期,日期一定合法且第二个日期一定晚于第一个日期。四个 0 表示输入结束,且不用处理。

输出:对每一行输入给定一行输出。输出两个日期间隔多少天。

样例输入:
```
    1 1 1 2
```
样例输出:

　　　　1

备注：样例输入表示1月1日和1月2日，间隔为1天。

1003　一元二次函数（题号 11620）

题目描述：很多人都说数学对计算机专业非常重要，例如现在给定一个一元二次函数及其定义域，求其极值。一元二次函数形如：y = ax^2+bx+c

输入：输入若干行，每行五个整数，即 a、b、c、x1、x2，每个数之间用一个空格分隔。a、b、c 的绝对值均不大于 10，[x1, x2]表示 x 的取值范围，x2 一定不小于 x1，且两者绝对值均不大于 100。连续五个 0 表示输入结束，且不用处理。

输出：每行输出两个数，用一个空格分隔，分别表示 y 的最小值与最大值（题目保证二次函数的对称轴及答案都是整数）。

样例输入：

　　1 0 0 0 1

样例输出：

　　0 1

备注：样例表示 y=x^2 在[0, 1]上的取值，所以 y 的最小值为 0，最大值为 1。

1004　数位的和（题号 11621）

题目描述：给定一个 32 位非负整数，求其所有位上的数字的和。

输入：若干行，每行一个整数。

输出：输出一行为答案。

样例输入：

　　34
　　450
　　56

样例输出：

　　7
　　9
　　11

1005　四位水仙花数（题号 11622）

题目描述：四位水仙花数是指一个四位数，满足如下条件：设该四位数形如 abcd，则 abcd=a^4+b^4+c^4+d^4。

输入：输入有若干行，每行一个 32 位正整数。

输出：对每个输入判断是否为四位水仙花数，是则输出 Y，否则输出 N。

样例输入：

　　1000

样例输出：

　　N

1006 次大公约数（题号 11623）

题目描述：给定两个 1000 以内的正整数，求其次大公约数。

输入：输入若干行，每行两个数，用一个空格分隔。

输出：对每个输入，输出一行为其答案。如果两个数是互质的，则输出 0。

样例输入：

    8 12
    8 15

样例输出：

    2
    0

上机实验考试模拟试卷（3）

1001 简单的题 2016（题号 11675）

题目描述：每一届的师兄师姐都会碰到一道简单的题，2016 级也不例外。而且，2016 级是有史以来最简单的题。因为今年是出题人参加高考 20 周年！嗯，年龄暴露题目。

输入：输入一行，为一个 32 位 int 类型的整数。

输出：输出一行，照输入原样输出。

样例输入：

    300

样例输出：

    300

1002 第三大的数（题号 11676）

题目描述：给 n 个数，求第三大的数。注意本题中第三大的定义。如果输入的三个数分别是 1、2、3，则答案为 1。如果输入的三个数分别是 3、2、2，则答案为 2。如果输入的三个数分别是 3、3、3，则答案为 3。

总之，按照数值从大到小进行排列，一样大的也要依次排列，最后处在第三个位置上的数就认为是第三大的数。

输入：首先是一个整数 n，0<n≤1000。然后是 n 个 32 位 int 类型的整数。

输出：输出一行，为第三大的数。如果少于三个数，则输出 NONE。

样例输入：

    2 2 1

样例输出：

    NONE

1003　字符串、数字分不清楚（题号 11677）

题目描述：给定一个字符串，问是否符合数字的格式。字符串的长度最多为 8 个。所谓合法的数字格式，就是指

（1）整数是合法的。整数就是由 0~9 十个数字组成。注意，我们认为正负号是不能包含在其中的，但是诸如 000 之类的数字是合法的。

（2）实数是合法的。实数就是带小数点的数。例如 3.1415926、2.71828 等都是合法的。特别地，由于是 C++，我们认为 3.和.3 也是合法的，其数值分别是 3.0 和 0.3。同样，正负号是不合法的。

输入：输入一行不带空格的字符串。

输出：如果是合法的数字字符串，则输出 YES，否则输出 NO。

样例输入：

    abcd

样例输出：

    NO

1004　挤空格（题号 11678）

题目描述：给定一个字符串，长度不超过 100 个字符，将其中的空格删掉，剩下的字符按原顺序输出。

输入：输入一行为一个字符串。

输出：输出为一个无空格的字符串。

样例输入：

    Hello, the world!

样例输出：

    Hello,theworld!

1005　两条直线（题号 11679）

题目描述：平面上有两条直线，确定其位置关系。

输入：输入一共八个整数，分别表示 A、B、C、D 四个点的坐标。

输出：如果直线 AB 与直线 CD 相交，则输出 INTER，否则输出 NO。

样例输入：

    0 0 1 1 1 0 0 1

样例输出：

    INTER

1006　上楼梯质数（题号 11680）

题目描述：我们称如下的数为上楼梯质数：13、239、569。首先它是质数，其次其数字从高位到低位是越来越大（不包含等于）的数。449 虽然是质数，但它不是上楼梯质数。

输入：输入只有一个正的 int 类型的数。

输出：如果是上楼梯质数，则输出 YES，否则输出 NO。

样例输入：

         13
样例输出：
         YES

笔试模拟试卷（1）参考答案

一、单选题（每题 2 分，共 30 分）
1~5 ①④③④②；6~10 ①④④②③；11~15 ②①③④②。

二、填空题（每空 1 分，共 12 分）
1. 5.5   2. 36   3. f   4. 0   6
5. strcat(s1,s2) 6. 51   7. a>=-10&&a<=10   8. 基类   派生类
9. 纯虚函数   10. else return n*Fun(n-1);

三、写出下列程序的运行结果（每题 4 分，共 16 分）
1.
51
60,21
2.
n=7
j=0
s=7
3.
Hello everybody!
Len=16
4.
1 2 3
8 9 4
7 6 5

四、程序改错（每题 4 分，共 12 分）
1. #include<iostream>
   using namespace std;
   void swap(int x,int y)   //改为 void swap(int &x,int &y)
   {
   int temp;

· 220 ·

```
        temp=x;
        x=y;
        y=temp;
    }
    int main()
    {
        int x=5,y=6;
        cout<<x<<" "<<y<<endl;
        swap(&x,&y);              //改为 swap(x,y);
        cout<<x<<" "<<y<<endl;
        return 0;
    }
2. #include<iostream>
   using namespace std;
   #define M 14
   #define N 7
   int main()
   {
        char a[M]="I Love China",b[N];
        int i,j=0;
        for(i=1;i<=M;i=i+2)    //改为 for(i=1;i<M;i=i+2)
        {
            if(a[i]>='a' && a[i]<='z')
            {
                b[j]=a[i]-32;
                j++;
            }
            else
            {
                b[j]=a[i];
                j++;
            }
        }
        cout<<"\n数组 b 为: "<<b[N];   //改为 cout<<"\n数组 b 为: "<<b;
        return 0;
   }
3. #include<iostream>
   using namespace std;
```

```
class Point
{
private:
    int x,y;
//插入public:
    void init(int a,int b){x=a;y=b;}      //改为Point(int a,int
                                          //       b){x=a;y=b;}
    void show(){cout<<"x=  "<<x<<"   y="<<y<<endl;}
};
int main()
{
    Point a(24,50);
    a.show();
    return 0;
}
```

五、程序填空（每空2分，共10分）

1. ①p1=p2=(struct student *)malloc(LEN);
   ②p2->next=p1;
   ③return(head);
2. ①return *this;
   ②cout<<x.getAmount()<<endl;

六、编程题（每题10分，共20分）

1.
```
#include<iostream>
using namespace std;
const int n=10;
void InsertSort(float a[],int n)
{
    int i,j;
    float x;
    for(i=1;i<n;i++)
    {
        x=a[i];//将待排序的元素a[i]存储在x中
        for(j=i-1;j>=0;j--)//寻找插入位置
            if(x>a[j])
                a[j+1]=a[j];//后移一个位置
            else
```

```cpp
                break;
            a[j+1]=x;//将x插入已找到的插入位置
        }
    }
    int fail_num(float a[],int n)
    {
        int i,num=0;
        for(i=0;i<n;i++)
            if(a[i]<60.0)
                num++;
        return num;
    }
    float avg_score(float a[],int n)
    {
        int i;
        float avg=0,sum=0;
        for(i=0;i<n;i++)
            sum+=a[i];
        avg=sum/n;
        return avg;
    }
    int main()
    {
        float a[n]={42,65,80,74,36,44,28,65,94,72};//定义一个数组
        int num;//不及格人数
        float avg;//平均分
        InsertSort(a,n);//调用函数进行插入排序
        for(int i=0;i<n;i++)//输出排序后的结果
            cout<<a[i]<<" ";
        cout<<endl;
        num=fail_num(a,n);//求学生不及格人数
        cout<<"不及格人数为: "<<num<<endl;
        avg=avg_score(a,n);//求平均成绩
        cout<<"平均成绩为: "<<avg<<endl;
        return 0;
    }
2. #include<iostream>
   using namespace std;
```

```cpp
const double pi=3.14;
class Shape
{ public:
    virtual void show_area()=0;
    virtual void show_perimeter()=0;
};
class Rectangle:public Shape
{    double side;
public:
    Rectangle(double s):side(s){}
    void show_area()
    {
        cout<<"Rectangle area is "<<side*side<<endl;
    }
    void show_perimeter()
    {
        cout<<"Rectangle perimeter is "<<side*4<<endl;
    }
};
class Circle:public Shape
{
    double radius;
public:
    Circle(double r):radius(r){}
    void show_area()
    {
        cout<<"Circle area is "<<pi*radius*radius<<endl;
    }
    void show_perimeter()
    {
        cout<<"Circle perimeter is "<<2*pi*radius<<endl;
    }
};
int main()
{
    Rectangle r(100);
    Circle c(100);
    r.show_area();
    r.show_perimeter();
```

```
    c.show_area();
    c.show_perimeter();
    return 0;
}
```

笔试模拟试卷（2）参考答案

一、单选题（每题 2 分，共 30 分）

1~5 ③④②②②；6~10 ③④③②①；11~15 ④①③②④。

二、填空题（每空 1 分，共 12 分）

1. 必须为字母或下划线　　2. −60　　3. 1　　4. 单继承　多继承
5. 4.5　　　　　　　　　6. 51　　　　　　　7. 直接递归　间接递归
8. 4　　　　　　　　　　9. public　　　　　　10. 15

三、写出下列程序的运行结果（每题 4 分，共 16 分）

1.
51
60,21
2.
 #
 # ##
#####
3.
1
5
5
4
3
2
1
1
4.
1 2
3

四、程序改错（每题4分，共12分）

1.
```cpp
#include<iostream>
using namespace std;
class base
{
protected:
    int i,j;
public:
    void set(int a,int b){i=a;j=b;};
    void show(){cout<<i<<" "<<j<<"\n";};
};
class derived1:private base   //修改public
{
    int k;
public:
    void setk(){k=i*j;};
    void showk(){cout<<k<<"\n";};
};
class derived2: public derived1
{
    int m;
public:
    void setm(){m=i-j;};
    void showm(){cout<<m<<"\n";};
};
void main()
{
    derived1 ob1;
    derived2 ob2;
    ob1.set(1,2);
    ob1.show();
    ob2.set(3,4);
    ob2.show();
    return 0;    //删去 或者将void main()改为int main()
}
```

2.
```cpp
#include<iostream>
using namespace std;
```

```
    int main()
    {
        int i,j,a[3][2]={1,2,3,4,5,6};
        int *p=a;         //修改 int *p=a[0];
        for(i=0;i<3;i++)
        {
            for(j=0;j<2;j++)
                cout<<" "<<*(*(p+i)+j); //采用指针方式访问p所指向的二维数组
                //修改 cout<<" "<<*(p+i*2+j);
            cout<<endl;
        }
        return 0;
    }
```

3.
```
#include<iostream>
using namespace std;
class one
{
public: //插入
    int a1,a2;
public:
    void initial(int x1,int x2)//修改 one(int x1,int x2)
    {
        a1=x1;
        a2=x2;
    }
};
int main()
{
    one data(2,3);
    cout<<data.a1<<endl;
    cout<<data.a2<<endl;
    return 0;
}
```

五、程序填空（每空2分，共10分）

1. ① str[i]=(char *)malloc(10);
   ② cout<<str[i]<<endl;
2. ① a[i]>b[j]   ② i<3   ③ j<5

六、编程题（每题 10 分，共 20 分）

1.
```cpp
#include<iostream>
using namespace std;
#define N 100
void encrypt(char str[])   //加密
{
    const int arr[7] ={4,9,6,2,8,7,3};
    int n,temp;
    for(n=0;n<int(strlen(str));n++)
    {
        temp=str[n]+arr[n%7];
        /* 注意C++一个字符只分配一个字节，最大只能为数字127，*/
        if(temp>122)
            temp=temp%122+31;  //请考虑这里为什么不用加32？
        str[n]=temp;
    }
}
void decryption(char str[])//解密
{
    const int arr[7] ={4,9,6,2,8,7,3};
    int n;
    for(n=0;n<int(strlen(str));n++)
    {
        str[n]=str[n]-arr[n%7];
        if(str[n]<32)
            str[n]=str[n]+91;
    }
}
int main()
{
    char str[N]="This moment will nap,you will have a dream;But this moment study,you will interpret a dream.zero";
    cout<<"加密前:"<<str<<endl;
    encrypt(str);
    cout<<"加密后:"<<str<<endl;
    decryption(str);
    cout<<"解密后:"<<str<<endl;
```

```
        return 0;
    }
2.  #include<iostream>
    using namespace std;
    class circle
    {
        double radius;
    public:
        circle(double r){radius=r;}
        double getarea(){return radius*radius*3.14;}
    };
    class table
    {
        double height;
    public:
        table(double h){height=h;}
        double getheight(){return height;}
    };
    class roundtable:public table,public circle
    {
        char *color;
    public:
        roundtable(double h,double r,char c[]):circle(r),table(h)
        {
            color=new char[strlen(c)+1];
            strcpy(color,c);
        }
        char *getcolor() {return color;}
    };
    int main()
    {
        roundtable rt(0.8,1.2,"黑色");
        cout<<"圆桌属性数据:"<<endl;
        cout<<"高度: "<<rt.getheight()<<"米"<<endl;
        cout<<"面积: "<<rt.getarea()<<"平方米"<<endl;
        cout<<"颜色: "<<rt.getcolor()<<endl;
        return 0;
    }
```

笔试模拟试卷（3）参考答案

一、单选题（每小题 2 分，共 40 分）

1~5 ①①④③③； 6~10 ④③③④①； 11~15 ③②④③①； 16~20 ③④②④①。

二、程序修改题（每小题 6 分，共 24 分。打星号的行有错误，请就近写出正确代码，每道题有且只有两处错误，每处错误仅涉及一行，其他行不得改动。）

题号	答案
1	a2=k/10;
	return i;
2	t += 1.0 / k;
	return t;
3	for(i=1;i<n;++i)
	t == 0
4	t = 0
	t = t / 10;

三、程序填空题（每小题 9 分，共 18 分。在画线处填入符合语法的内容以完成函数功能。）

题号	第 1 空	第 2 空	第 3 空
1	10	0	x
2	0	n	ave

四、程序设计题（每小题 9 分，共 18 分）

1. 其他写法酌情计分。

```
int main(){           //main        1分
  int a,b,c;          //定义         2分
  cin>>a>>b>>c;       //输入         2分
  int r = a;
  if ( r > b ) r = b;
  if ( r > c ) r = c;  //运算        2分
  cout<<r<<endl;      //输出         2分
  return 0;
}
```

2. 其他写法酌情计分。

```
int main(){           //main        1分
```

```
    int r = 0;
    string s;                    //定义           2分
    cin >> s;                    //输入           2分
    for(int i=0;i<s.length();++i){
        if ( islower(s[i]) ) ++r;
    }                            //运算           2分
    cout<<r<<endl;               //输出           2分
    return 0;
}
```

笔试模拟试卷（4）参考答案

一、单选题（每小题 2 分，共 40 分）

1~5 ①①④③③；6~10 ④③③②①；11~15 ③②③①③；16~20 ③④②④③。

二、程序修改题（每小题 6 分，共 24 分。打星号的行有错误，请就近写出正确代码，每道题有且只有两处错误，每处错误仅涉及一行，其他行不得进行改动。）

题号	答案
1	*s != '\0'
	++s;
2	s[j++]=s[i];
	'\0'
3	t = 1;
	return s
4	i <= m
	y += 1.0 / (i * i);

三、程序填空题（每小题 9 分，共 18 分。在画线处填入符合语法的内容以完成函数功能。）

题号	第1空	第2空	第3空
1	s	--	true
2	0	x[i]	s

四、程序设计题（每小题 9 分，共 18 分）

1. 其他写法酌情计分
```
    int main(){              //main            1分
```

```
   int a,b,c;              //定义       2分
   cin>>a>>b>>c;           //输入       2分
   int r = a;
   if ( r < b ) r = b;
   if ( r < c ) r = c;     //运算       2分
   cout<<r<<endl;          //输出       2分
   return 0;
}
```

2. 其他写法酌情计分
```
   int main(){             //main      1分
   int r = 0;
   string s;               //定义       2分
   cin >> s;               //输入       2分
   for(int i=0;i<s.length();++i){
       if ( isupper(s[i]) ) ++r;
   }                       //运算       2分
   cout<<r<<endl;          //输出       2分
   return 0;
}
```

笔试模拟试卷（5）参考答案

一、单选题（每小题 2 分，共 40 分）

1~5 ①③③②③；6~10 ①②③②④；11~15 ③①①①③；16~20 ③②②②①。

二、程序修改题（每小题 6 分，共 24 分。打星号的行有错误，请就近写出正确代码，每道题有且只有两处错误，每处错误仅涉及一行，其他行不得改动。）

题号	答案
1	int k = 1;
	num /= 10;
2	char *a
	cout<<*a<<endl;
3	不要分号
	分号改成冒号
4	s[j++]=s[i];
	s[j]='\0';

三、程序填空题（每小题 9 分，共 18 分。在画线处填入符合语法的内容以完成函数功能。）

题号	第 1 空	第 2 空	第 3 空
1	n%10	break	2
2	x[0]	<	ret

四、程序设计题（每小题 9 分，共 18 分）

1. 其他写法酌情计分

```
int main(){            //main        1分
    int a,b,c;         //定义        2分
    cin>>a>>b>>c;      //输入        2分
    int r = a;
    if ( r < b ) r = b;
    if ( r < c ) r = c;
    int r2 = a;
    if ( r2 > b ) r2 = b;
    if ( r2 > c ) r2 = c;   //运算    2分
    cout<<a+b+c-r-r2<<endl; //输出    2分
    return 0;
}
```

2. 其他写法酌情计分

```
int main(){                         //main    1分
    int r = 0;
    string s;                       //定义    2分
    cin >> s;                       //输入    2分
    for(int i=0;i<s.length();++i){
        if ( isdigit(s[i]) ) ++r;
    }                               //运算    2分
    cout<<r<<endl;                  //输出    2分
    return 0;
}
```

上机实验考试模拟试卷（1）参考答案

1001

```
#include <iostream>
using namespace std;
typedef long long int llt;
```

```cpp
int main(){
    int a,b;
    while( 1 ){
            cin>>a>>b;
            if ( 0 == a && 0 == b ) return 0;
            cout<<((llt)a+(llt)b))<<endl;
    }
    return 0;
}
```

1002

```cpp
#include <iostream>
using namespace std;

inline int isLeap(int year){
    return ( 0 == year % 400 ||
    ( 0 == year % 4 && year % 100 ) )
    ? 1 : 0;
}
int A[] = {0,31,28,31,30,31,30,31,31,30,31,30,31};
int main(){
    int y,m,d1,d2;
    while( cin >> y >> m >> d1 >> d2) ){
        int ret = A[m] - d1 + 1;
        ret += d2 + (( 2 == m ) ? isLeap(y) : 0);
        cout << ret << endl;
    }
    return 0;
}
```

1003

```cpp
#include <iostream>
using namespace std;
int A[1001];
int main(){
    int n;
    while(cin>>n){
        for(int i=0;i<n;++i) cin>>A[i];

        int m = A[0];
```

```cpp
        for(int i=1;i<n;++i)
            if ( A[i] > m )
                m = A[i];

        int r = -1;
        for(int i=0;i<n;++i){
            if ( A[i] == m ) continue;
            if ( A[i] > r ) r = A[i];
        }
        cout<<r<<endl;
    }
    return 0;
}
```

1004

```cpp
#include <iostream>
using namespace std;

bool isPrime(int n){
    if ( 1 == n ) return false;
    for(int i=2;i*i<=n;++i)
        if ( 0 == n % i )
            return false;
    return true;
}
int sum(int a,int b){
    int s = 0;
    for(int i=a;i<=b;++i)
        if ( isPrime(i) )
            s += i;
    return s;
}
int main(){
    int a,b;
    while( cin>>a>>b ){
        cout<<sum(a,b)<<endl;
    }
    return 0;
}
```

上机实验考试模拟试卷（2）参考答案

1001

```cpp
#include <iostream>
using namespace std;
typedef long long int llt;

int main(){
    int a,b;
    while( 1 ){
        cin>>a>>b;
        if ( 0 == a && 0 == b ) return 0;
        cout<<((llt)a-(llt)b)<<endl;
    }
    return 0;
}
```

1002

```cpp
#include <iostream>
using namespace std;

int A[] = {0,31,28,31,30,31,30,31,31,30,31,30,31};

int f(int m1,int d1,int m2,int d2){
    if ( m1 == m2 ) return d2 - d1;

    int r = 0;
    for(int i=m1+1;i<m2;++i)  r += A[i];
    return r + d2 + A[m1] - d1;
}

int main(){
    int m1,d1,m2,d2;
    while( 1 ) {
        cin >> m1 >> d1 >> m2 >> d2;
```

```cpp
            if ( 0 == m1 ) return 0;
            cout<<f(m1,d1,m2,d2)<<endl;
        }
        return 0;
    }
```

1003

```cpp
#include <iostream>
using namespace std;

int max(int a,int b){
    return a > b ? a : b;
}

int max(int a,int b,int c){
    return max(a,max(b,c));
}

int min(int a,int b){
    return a < b ? a : b;
}

int min(int a,int b,int c){
    return min(a,min(b,c));
}

int f(int b,int c,int x){
    return b * x + c;
}

int f(int a,int b,int c,int x){
    return a * x * x + b * x + c;
}

int main(){
    int a,b,c,x1,x2;
    while(1){
        cin>>a>>b>>c>>x1>>x2;
```

```
        if ( 0 == a && 0 == b && 0 == c && 0 == x1 && 0 == x2 ) return 0;
        if ( 0 == a ){
            int y1 = f(b,c,x1);
            int y2 = f(b,c,x2);
            if ( b >= 0 ) cout<<y1<<" "<<y2<<endl;
            else cout<<y2<<" "<<y1<<endl;
            continue;
        }

        int x0 = - b / 2 / a;
        int y1 = f(a,b,c,x1);
        int y2 = f(a,b,c,x2);
        int y0 = f(a,b,c,x0);

        if ( a > 0 ){
            if ( x1 <= x0 && x0 <= x2 ){
                cout<<y0<<" "<<max(y1,y2)<<endl;
            }else{
                cout<<min(y1,y2)<<" "<<max(y1,y2)<<endl;
            }
        }else{
            if ( x1 <= x0 && x0 <= x2 ){
                cout<<min(y1,y2)<<" "<<y0<<endl;
            }else{
                cout<<min(y1,y2)<<" "<<max(y1,y2)<<endl;
            }
        }

    }

    return 0;
}
```

1004

```
#include <iostream>
using namespace std;
```

```
    int f(int n){
        int r = 0;
        while(n){
            r += n % 10;
            n /= 10;
        }
        return r;
    }
    int main(){
        int n;
        while( cin>>n ){
            cout<<f(n)<<endl;
        }
        return 0;
    }
```

1005

```
    #include <iostream>
    using namespace std;


    int f(int n){
        int r = 0;
        while(n){
            int x = n % 10;
            r += x * x * x * x;
            n /= 10;
        }
        return r;
    }

    int main(){
        int n;
        while( cin>>n ){
            if ( n < 1000 || n > 9999 ){
                cout << "N" << endl;
            }else{
                cout<<( f(n) == n ? "Y" : "N" )<<endl;
```

```
        }
    }
    return 0;
}
```

1006

```cpp
#include <iostream>
using namespace std;

int main(){
    int a,b;
    while( cin>>a>>b ){
        int r = a < b ? a : b;
        int t = 0;
        for(;r>=1;--r){
            if ( 0 == a % r && 0 == b % r ){
                ++t;
            }
            if ( 2 == t ) break;
        }
        cout<<r<<endl;
    }
    return 0;
}
```

上机实验考试模拟试卷（3）参考答案

1001

```cpp
#include <iostream>
using namespace std;

int main(){
    int a;
    cin >> a;
    cout << a << endl;
```

```
    return 0;
}
```

1002

```cpp
#include <iostream>
using namespace std;

int A[1100];

int f(int a[],int n){
    int r = 0;
    for(int i=1;i<n;++i){
        if ( a[r] < a[i] ) r = i;
    }
    return r;
}

int main(){
    int n;
    cin >> n;
    for(int i=0;i<n;++i) cin >> A[i];

    if ( n < 3 ){
        cout<<"NONE"<<endl;
        return 0;
    }

    int m1 = f(A,n);
    A[m1] = A[n-1];

    int m2 = f(A,n-1);
    A[m2] = A[n-2];

    int m3 = f(A,n-2);
    cout<<A[m3]<<endl;
    return 0;
}
```

1003

```cpp
#include <iostream>
#include <string>
#include <cctype>
using namespace std;

bool isOK(string &s){
    if ( s == "." ) return false;
    int cnt = 0;
    for(int i=0;i<s.length();++i){
        if ( isdigit(s[i]) ){
            ;
        }else if ( '.' == s[i] ){
            ++cnt;
            if ( cnt > 1 ) return false;
        }else{
            return false;
        }
    }
    return true;
}
int main(){
    string s;
    cin >> s;
    cout<<(isOK(s)?"YES":"NO")<<endl;
    return 0;
}
```

1004

```cpp
#include <iostream>
#include <string>
using namespace std;

int main(){
    string s;
    getline(cin,s);
    for(int i=0;i<s.length();++i){
```

```
        if ( ' ' != s[i] ) cout<<s[i];
    }
    cout<<endl;
    return 0;
}
```

1005

```
#include <iostream>
using namespace std;

struct T{
    int x,y,t;
    T(int xx=0,int yy=0,int tt=0):x(xx),y(yy),t(tt){}
};
T f(T const&A,T const&B){
    return T(A.y*B.t-B.y*A.t,B.x*A.t-A.x*B.t,A.x*B.y-B.x*A.y);
}
int main(){
    T A,B,C,D;
    cin>>A.x>>A.y>>B.x>>B.y>>C.x>>C.y>>D.x>>D.y;
    A.t=B.t=C.t=D.t=1;

    T AB = f(A,B);
    T CD = f(C,D);

    T ABCD = f(AB,CD);

    if ( ABCD.t || ( 0 == ABCD.x && 0 == ABCD.y && 0 == ABCD.t ) )
        cout<<"INTER"<<endl;
    else cout<<"NONE"<<endl;

    return 0;
}
```

1006

```
#include <iostream>
#include <string>
using namespace std;
```

```cpp
bool isPrime(int n){
    for(int i=2;i<n/2;++i)if(n%i==0)return false;
    return true;
}
bool isUp(int n){
    int a[20] = {n % 10};
    n /= 10;
    int k = 1;
    while(n){
        a[k] = n % 10;
        if ( a[k] >= a[k-1] ) return false;
        n /= 10;
        ++k;
    }
    return true;
}
int main(){
    int n;
    cin>>n;
    if ( isPrime(n) && isUp(n) ) cout<<"YES"<<endl;
    else cout<<"NO"<<endl;
    return 0;
}
```

附 录

附录 A ASCII 码对照表

ASCII 码对照表如下：

ASCII 值	控制字符	ASCII 值	控制字符	ASCII 值	控制字符	ASCII 值	控制字符
0	NUL	32	(space)	64	@	96	、
1	SOH	33	!	65	A	97	a
2	STX	34	"	66	B	98	b
3	ETX	35	#	67	C	99	c
4	EOT	36	$	68	D	100	d
5	ENQ	37	%	69	E	101	e
6	ACK	38	&	70	F	102	f
7	BEL	39	,	71	G	103	g
8	BS	40	(	72	H	104	h
9	HT	41	)	73	I	105	i
10	LF	42	*	74	J	106	j
11	VT	43	+	75	K	107	k
12	FF	44	,	76	L	108	l
13	CR	45	-	77	M	109	m
14	SO	46	.	78	N	110	n
15	SI	47	/	79	O	111	o
16	DLE	48	0	80	P	112	p
17	DC1	49	1	81	Q	113	q
18	DC2	50	2	82	R	114	r
19	DC3	51	3	83	S	115	s
20	DC4	52	4	84	T	116	t
21	NAK	53	5	85	U	117	u
22	SYN	54	6	86	V	118	v
23	TB	55	7	87	W	119	w
24	CAN	56	8	88	X	120	x
25	EM	57	9	89	Y	121	y
26	SUB	58	:	90	Z	122	z
27	ESC	59	;	91	[	123	{
28	FS	60	<	92	\	124		
29	GS	61	=	93	]	125	}
30	RS	62	>	94	^	126	~
31	US	63	?	95	—	127	DEL

附录 B C/C++ 与标准 C++ 头文件对照表

C/C++ 与标准 C++ 头文件对照表如下：

传统 C/C++	标准 C++	作用
#include <ctype.h>	#include <cctype>	字符处理
#include <errno.h>	#include <cerrno>	定义错误码
#include <locale.h>	#include <clocale>	定义本地化函数
#include <math.h>	#include <cmath>	定义数学函数
#include <stdio.h>	#include <cstdio>	定义输入／输出函数
#include <stdlib.h>	#include <cstdlib>	定义杂项函数及内存分配函数
#include <string.h>	#include <cstring>	字符串处理
#include <time.h>	#include <ctime>	定义关于时间的函数
#include <limits.h>	#include <limits>	定义整型和字符型数据类型的限制值
#include <iomanip.h>	#include <iomanip>	参数化输入／输出
#include <wchar.h>	#include <cwchar>	宽字符处理及输入／输出
#include <wctype.h>	#include <cwctype>	宽字符分类
#include <iostream.h>	#include <iostream>	数据流输入／输出
#include <fstream.h>	#include <fstream>	文件输入／输出
	#include <algorithm>	STL 通用算法
	#include <bitset>	STL 位集容器
	#include <complex>	复数类
	#include <deque>	STL 双端队列容器
	#include <exception>	异常处理类
	#include <functional>	STL 定义运算函数（代替运算符）
	#include <list>	STL 线性列表容器
	#include <map>	STL 映射容器
	#include <ios>	基本输入／输出支持
	#include <iosfwd>	输入／输出系统使用的前置声明
	#include <istream>	基本输入流
	#include <ostream>	基本输出流
	#include <queue>	STL 队列容器
	#include <set>	STL 集合容器
	#include <sstream>	基于字符串的流
	#include <stack>	STL 堆栈容器
	#include <stdexcept>	标准异常类
	#include <streambuf>	底层输入／输出支持
	#include <string>	字符串类
	#include <utility>	STL 通用模板类
	#include <vector>	STL 态数组容器

在标准 C++ 的包含头文件后，需要使用命名空间 "using namespace std;"。

附录 C　Linux、UNIX 下编译 C++ 程序

ACM 国际大学生程序设计竞赛的区域赛和全球总决赛的比赛环境一般都是 Linux 或 UNIX 等操作系统，因此，作为参赛选手，必须熟悉在这些操作系统下如何编译和运行 C++ 程序。下面就 Linux 下编译 C++ 程序作简要介绍。

GCC（GNU Compiler Collection，GNU 编译器套装），是一套由 GNU 开发的编程语言编译器。它是一套以 GPL 及 LGPL 许可证发行的自由软件，是 GNU 计划的关键部分，也是自由的类 UNIX 及苹果计算机 Mac OS X 操作系统的标准编译器。GCC（特别是其中的 C 语言编译器）也常被认为是跨平台编译器的事实标准。

GCC 原名为 GNU C 语言编译器（GNU C Compiler），因为它原本只能处理 C 语言。GCC 很快地扩展，变得可处理 C++ 语言。之后也变得可处理 Fortran、Pascal、Objective-C、Java，以及 Ada 与其他语言。

下面介绍两个实例，分别如附例 C.1、附例 C.2 所示。

【附例 C.1】helloworld.cpp。

```
#include<iostream>
using namespace std;
int main()
{
    cout<<"hello, world"<<endl;
    return 0;
}
```

1. 单个源文件生成可执行程序

（1）`$ g++ helloworld.cpp`

由于命令行中未指定可执行程序的文件名，所以编译器采用默认的 a.out。

运行：`$ ./a.out`

（2）`$ g++ helloworld.cpp -o helloworld`

这里是通过 -o 选项指定可执行程序的文件名。上面的命令将产生名为 helloworld 的可执行文件。

运行：`$ ./helloworld`

（3）`$ gcc helloworld.cpp -lstdc++ -o helloworld`

程序 g++ 将 gcc 默认语言设为 C++ 的一个特殊版本，连接时它自动使用 C++ 标准库而不使用 C 标准库。选项 -l (ell) 通过添加前缀 lib 和后缀 .a 将跟随它的名字变换为库的名字 libstdc++.a，而后 -l (ell) 在标准库路径中查找该库。gcc 的编译过程和输出文件与 g++ 是完全相同的。

在大多数系统中，gcc 安装时会安装一个名为 C++ 的程序。如果被安装，则它与 g++

等同。
```
$ c++ helloworld.cpp -o helloworld
```
运行：`$ ./ helloworld`

2. 多个源文件生成可执行程序

【附例 C.2】
```cpp
/* speak.h */
#include <iostream>
class Speak
{
    public:
        void sayHello(const char *);
};

/* speak.cpp */
#include "speak.h"
void Speak::sayHello(const char *str)
{
    std::cout << "Hello " << str << "\n";
}

/* hellospeak.cpp */
#include "speak.h"
int main(int argc,char *argv[])
{
    Speak speak;
    speak.sayHello("world");
    return 0;
}
```

下面这条命令将上述两个源文件编译链接成一个单一的可执行程序 hellospeak：
```
$ g++ hellospeak.cpp speak.cpp -o hellospeak
```

3. 源文件生成对象文件

```
$ g++ -c helloworld.cpp
```

选项-c 用来告诉编译器编译源码但不要执行链接，输出结果为对象文件。文件默认名与源码文件名相同，只是将其后缀变为.o。

选项-o 不仅仅能用来命名可执行文件，它也可用来命名编译器输出的其他文件。

生成对象文件后需要生成可执行文件才能执行，如：
```
$ g++ -o helloworld.o -o helloworld
```

4. 编译预处理

```
$ g++ -E helloworld.cpp
```

选项-E 使 g++将源码用编译预处理器处理后不再执行其他动作。上面的命令预处理源码文件 helloworld.cpp 并将结果显示在标准输出中。

5. 生成汇编代码

```
$ g++ -S helloworld.cpp
```

选项-S 用来指示编译器将程序编译成汇编语言，输出汇编语言代码而后结束。上面的命令将由 C++源码文件生成汇编语言文件 helloworld.s。生成的汇编语言依赖于编译器的目标平台。

6. 其他部分编译选项

-w 表示关闭所有警告。

-Wall 表示允许所有有用的警告。

-g 表示生成的目标文件中带调试信息，gdb 能够使用这些调试信息。

在学习时，如果没有 linux 环境，也可以在 Dev-C++中模拟使用下面的命令。方法如下：

（1）启动命令提示符窗口。依次选择"开始"→"运行"，输入"CMD"，回车进入。

（2）进入 Dev-C++的 bin 目录下。下面假设 Dev-C++安装在 C 盘根目录下，编译 C:\test.cpp。过程如附图 C.1 所示。

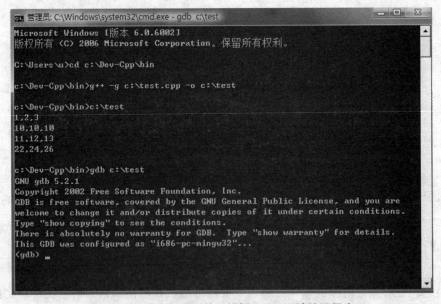

附图 C.1 命令行提示符下模拟 Linux 环境编译程序

7. gdb 调试

（1）如果要对程序进行调试，编译程序时需要添加-g 选项，将调试信息添加到可执行文件中。如：

```
$ g++ -g test.cpp -o test
```

（2）启动 gdb——gdb <program>

```
$ gdb test
```

（3）在 GDB 中运行程序——run。

（4）其他部分命令如附表 C.1 所示。

附表 C.1　部分命令及其作用

命令	作用
list <linenum>	显示程序第 linenum 行周围的源程序
list	显示当前行后面的源程序
break <linenum>	设置断点
info break [n]	查看断点
watch <expr>	设置观察点
info watchpoints	列出当前设置了的所有观察点
clear	清除所有已定义的停止点
clear <linenum>	清除所有设置在指定行上的停止点
delete [breakpoints] [range...]	删除指定的断点
disable [breakpoints] [range...]	breakpoints 为停止点号。disable 的停止点，GDB 不会删除，当你还需要时，enable 即可
enable [breakpoints] [range...]	enable 所指定的停止点，breakpoints 为停止点号
continue [ignore-count]	恢复程序运行，直到程序结束，或是下一个断点到来。ignore-count 表示忽略其后的断点次数
step <count>	单步跟踪，如果有函数调用，则进入该函数。进入函数的前提是此函数被编译为有 debug 信息。不添加 count，表示一条条地执行；添加 count，表示执行后面的 count 条指令，然后再停住
next <count>	同样单步跟踪，如果有函数调用，则不会进入该函数
finish	运行程序，直到当前函数完成返回，并打印函数返回时的堆栈地址、返回值及参数值等信息
print <expr>	查看运行时的数据
backtrace	打印当前的函数调用栈的所有信息
backtrace <n>	只打印栈顶上 n 层的栈信息
info registers	查看寄存器的情况
display <expr>	当程序停住时，或是在你单步跟踪时，这些变量会自动显示
disable display <dnums...>	让自动显示失效
enable display <dnums...>	让自动显示恢复
delete display <dnums...>	删除自动显示
jump <linespec>	指定下一条语句的运行点
q	退出 gdb

更多命令的使用请参考 gdb 帮助。

附录D 在Visual C++下调试程序

1. 调试概述

调试是编程过程中不可忽视的问题,对于一个复杂的程序来说,从编写程序到能够通过编译,只是完成了一小部分而已。然后还要不断地调试、修改、再调试、再修改……直到将发现的问题都解决,程序能够稳定正确运行为止。调试在程序设计中的作用是十分重要的。一段程序完成的速度和质量,往往取决于程序员的调试水平。如何准确定位产生错误的代码,是一项既需要技巧又需要经验的工作。

市场上有很多调试产品供选择,不同的调试工具面向的应用是不同的。Visual C++ 6的 IDE 为我们内置了一套非常优秀的调试工具,通过这套调试工具,我们可以直接在C/C++源代码级上调试程序,同时也可以在汇编代码级上调试程序,对于大部分应用来说,这已经足够了。

2. 调试环境的设置

一般来说,一个新建立的 VC 项目有两个版本:Debug 版本和 Release 版本。Debug版本是通常所说的调试版本。调试版本在编译生成的二进制文件中含有调试信息,因此通常文件会比较大。而 Release 版本就是通常所说的发布版本,通常文件会比较小。由于发布版本编译器会对程序进行优化,编译生成的代码可能与实际的代码并不一致,因此并不适合调试(可能出现断点错位的情况)。

对当前编译版本的设置可以通过菜单项中的"Build"→"Set Active Project Configuration"来设置。这里会显示当前 Workspace 中所有项目允许的编译版本,如附图D.1 所示。

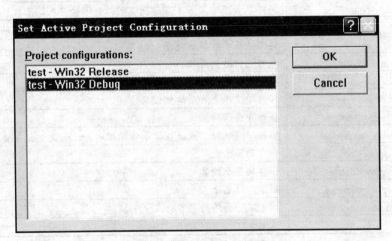

附图 D.1 编译版本的设置

在"Project"→"Setting…"中有一些调试相关的设置,这里简要介绍一下。

1）Debug 选项卡

Debug 选项卡如附图 D.2 所示。

左边的"Settings For:"表示当前的设置对哪一个编译版本有效。也就是说，可以对 Debug 版本和 Release 版本分别进行项目设置。

右边的"Executable for debug session:"表示当前项目调试的二进制文件。

"Working directory:"表示程序调试时的工作路径，默认设置为项目文件（扩展名为.dsp 的文件）所在目录，如果感觉不方便，就可以在这里设置工作路径。

"Program arguments:"表示启动程序的参数。

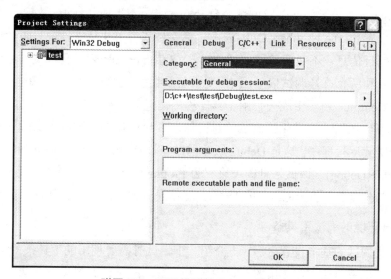

附图 D.2　项目设置的 Debug 选项卡

2）C/C++选项卡

C/C++选项卡如附图 D.3 所示。

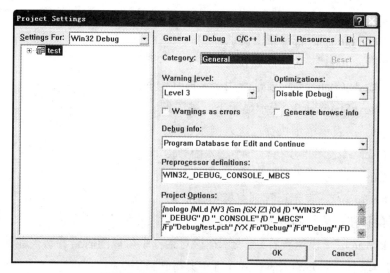

附图 D.3　项目设置的 C/C++选项卡

"Warning level:"表示编译时警告的等级。选项从 None 到 Level 4，None 表示没有警告信息，Level 4 表示警告最严格，通常使用默认设置即可。

"Optimizations:"表示编译优化设置。选项 Default 表示使用默认优化设置；Disable(Debug)表示禁止优化，这是 Debug 版本的默认设置；Maximize Speed 表示优化程序运行速度，这是 Release 版本的默认设置；Minimize Size 表示优化程序的大小，如果选择 Customize，则根据用户在【Category:】【Optimizations:】中设置的优化条件来优化编译。

"Debug info:"表示如何生成调试信息。None 表示不生成任何调试信息，这是 Release 版本的默认设置。Line Numbers Only 表示目标文件或者可执行文件中只包含全局和导出符号以及代码行信息，不包含符号调试信息。C7 Compatible 表示目标文件或者可执行文件中包含行号和所有符号调试信息，包括变量名及类型、函数及原型等。Program Database for Edit and Continue 表示生成一个用于存储程序信息的数据文件（扩展名为.pdb），这个文件包含类型信息以及符号化的调试信息。该选项允许在调试过程中修改代码或者变量取值并编译继续运行程序，这是 Debug 版本的默认设置。如果将 Release 版本的优化选项设置为 Disable(Debug)，调试信息选项设置为 Program Database for Edit and Continue，并且在"Link"选项卡中选上 Generate Debug info，则 Release 版本也可以调试了。

3．调试环境的调试工具条

在了解了调试的基本设置之后，再介绍一下调试时需要使用的工具条（见附图 D.4）。右键单击工具栏空白处，在弹出的菜单中选中"Debug"。任意建立一个基于 Console 的 VC 项目，在代码中通过 F9 键添加/取消断点，然后按 F5 键编译进入调试运行状态，当程序暂停在断点处时，Debug 工具条会呈现激活状态，下面介绍该工具条的作用。

第二排的按钮，从左到右依次如下。

"Restart"：表示重新调试运行。

"Stop Debugging"：表示停止调试。

"Break Execution"：表示断点执行。

"Apply Code Changes"：表示应用调试时代码的改变。

"Show Next Statment"：表示显示下一条语句。

"Step Into"：表示单步执行，进入函数内部。

"Step Over"：表示单步执行，不进入函数内部。

"Step Out"：表示跳出当前函数。

"Run to Cursor"：表示运行到鼠标指针处。

第二排的按钮，从左到右依次如下。

"Quick Watch"（快速查看）：表示在该对话框中可以随时查看感兴趣的变量以及函数，甚至其他信息，例如输入表达式 1+1，会得到结果 2。

"Watch"（查看）：表示这里可以查看各种信息，如变量、数值、错误信息等。而且可以设置输出的格式。

"Variables"（变量）：表示这里可以查看当前函数内的所有变量信息。Context 可以选择函数范围。

"Registers"（寄存器）：表示这里可以查看寄存器的当前状态。有时无法定位异常代码位置的时候，可以通过查看寄存器来判断可能发生的错误，这需要对寄存器的状态标志位有一些了解。

"Memory"（内存）：表示这里可以查看一片连续的内存。例如有时需要同时观察一个数组的某一个区域，可以直接查看该区域对应的内存。

"Call Stack"（调用堆栈）：表示这里可以查看函数的调用关系。对于比较复杂的应用程序，可以通过调用堆栈来跟踪程序运行的轨迹。

"Disassembly"（禁止汇编）：表示这里可以控制当前调试 C/C++源码或汇编源码。

另外，在 VC IDE 的最底部有一个调试信息输出窗口，这也是很有用的。包括 TRACE 的输出以及内存泄漏情况。

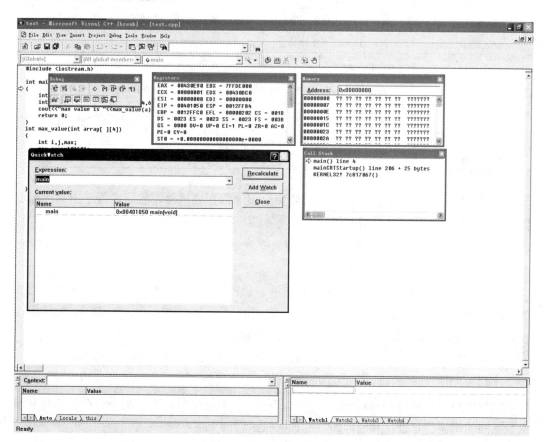

附图 D.4　调试工具条

附表 D.1 所示为调试时最常用的快捷键。希望读者在初学的时候尽量使用这些快捷键，熟练后将大大提高调试的效率。

附表 D.1 常用快捷键

快捷键	对应菜单功能	功能描述
F5	Debug→go	开始调试运行，继续调试运行
Ctrl+Shift+F5	Debug→Restart	重新调试运行
Shift+F5	Debug→Stop Debug	停止调试
F9		添加/删除断点
Ctrl+F9		允许/禁止断点
Shift+F9		快速查看窗口
Ctrl+Shift+F9		删除所有断点
Ctrl+b/Alt+F9	Edit→Break points	断点设置对话框
F10	Debug→Step Over	单步执行，不进入函数内部
F11	Debug→Step Into	单步执行，进入函数内部
Shift+F11	Debug→Step Out	跳出当前函数
Ctrl+F10	Debug→Run to Cursor	运行到鼠标指针处
Alt+F10	Debug→Apply Code Changes	应用调试时代码的改变
Alt+F11		切换 Memory 窗口显示格式，可以以不同字节格式来显示
Alt+Shift+F11		切换 Memory 窗口显示格式

一个好的程序员不应该把所有的判断交给编译器和调试器，应该在程序中自己加以程序保护和错误定位，具体措施包括以下几方面。

（1）对于所有有返回值的函数，都应该检查返回值，除非你确信这个函数调用绝对不会出错，或者不关心它是否出错。

（2）一些函数返回错误，需要用其他函数获得错误的具体信息。例如，accept 返回 INVALID_SOCKET，表示 accept 失败，为了查明具体的失败原因，应该立刻使用 WSAGetLastError 获得错误码，并有针对性地解决问题。

（3）有些函数通过异常机制抛出错误，应该用 TRY-CATCH 语句来检查错误

（4）对于能处理的错误，程序员应该自己在底层处理；对于不能处理的错误，程序员应该报告给用户让他们自己决定怎么处理。如果程序出现异常，却不对返回值和其他机制返回的错误信息进行判断，那只能加大找错误的难度。

对于 Visual Studio 2010 的调试与 Visual C++ 6.0 类似，因篇幅原因不再单独介绍。

附录 E　Dev-C++调试

1. 设置编译器选项

Dev-C++需要将编译器选项中的"产生调试信息"设置为"Yes"，才能进行程序的调试。方法如下。

选择 Tools（工具）→Compiler Options（编译器选项）→Settings（代码生成/优化）→Linker（连接器）→Generate Debugging Information（产生调式信息）→Yes，如附图 E.1 所示。

附图 E.1　编译器选项设置

2. 编译程序

请参考第 1 章。

3. 设置断点（Break Point）

将光标移动到想要暂停执行的那一行，按 Ctrl + F5 快捷键，或者直接用鼠标点击附图 E.2 中红线标明的区域，或者单击鼠标右键选择"切换断点"。

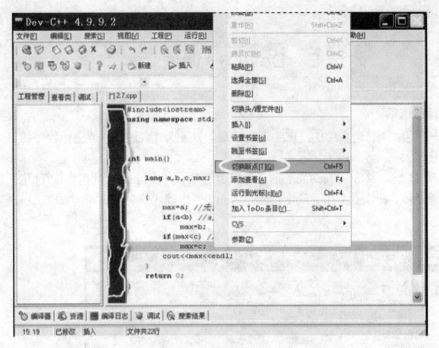

附图 E.2　设置断点

4. 调试（Debug）

按 F8 键开始调试。如果没有把"产生调试信息"设置为"Yes"，Dev-C++会提示工程没有调试信息，如附图 E.3 所示。

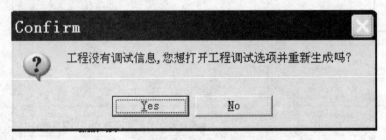

附图 E.3　Confirm 窗口

点击"Yes"，Dev-C++会自动把"产生调试信息"设置为"Yes"，并且重新编译工程。程序运行到断点处会暂停，如附图 E.4 所示。

按 F7 键执行当前行，并跳到下一行。

按 Ctrl＋F7 快捷键跳到下一断点。

5. 查看变量的值

开始调试后，在图示区域按右键，选择"Add Watch（添加查看）"；或者直接按 F4 键。在弹出的窗口中输入想查看的变量名，然后单击"确定"按钮，就可以看到该变量的值，分别如附图 E.5 和附图 E.6 所示。

附 录

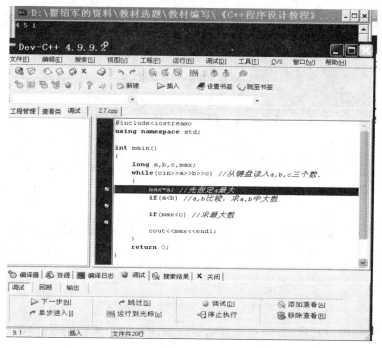

附图 E.4 调试窗口

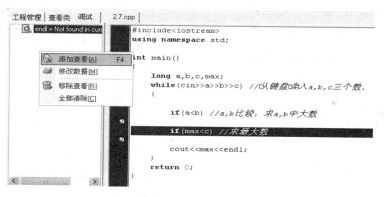

附图 E.5 添加查看

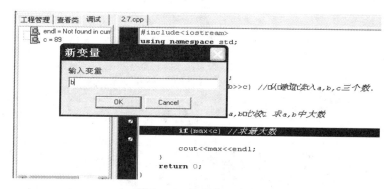

附图 E.6 输入查看变量

用鼠标选择源文件中的变量名，然后按 F4 也可以查看变量的值，该变量会出现在左边的调试列表中，如附图 E.7 所示。

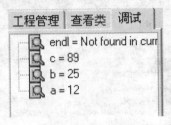

附图 E.7　调试列表

如果在 Environment Options（环境选项）中选择了"查看鼠标指向的变量（Watch Variable Under Mouse）"，用鼠标指向要查看的变量一段时间，该变量也会被添加到调试列表中。

提示：

（1）当想查看指针指向的变量的值时，按 F4 键，然后输入*及指针的名字（如 *pointer）。如果没加*，看到的将会是一个地址，也就是指针的值。

（2）有时调试器（Debugger）可能不知道某个指针的类型，从而不能显示该指针指向的变量的值。此时，需要手动输入该指针的类型。按 F4 键后，以*(type *)pointer 形式输入。例如，*(int *)pointer。